THE CHAO PHYA

THE CHAO PHYA
River in Transition

Steve Van Beek

Kuala Lumpur
OXFORD UNIVERSITY PRESS
Oxford Singapore New York
1995

Oxford University Press

Oxford New York

Athens Auckland Bangkok Bombay
Calcutta Cape Town Dar es Salaam Delhi
Florence Hong Kong Istanbul Karachi
Kuala Lumpur Madras Madrid Melbourne
Mexico City Nairobi Paris Singapore
Taipei Tokyo Toronto

and associated companies in

Berlin Ibadan

Oxford is a trade mark of Oxford University Press

Published in the United States
by Oxford University Press, New York

© *Oxford University Press 1995*
First published 1995

All rights reserved. No part of this publication may be reproduced, stored in a retrieval system, or transmitted, in any form or by any means, without the prior permission in writing of Oxford University Press. Within Malaysia, exceptions are allowed in respect of any fair dealing for the purpose of research or private study, or criticism or review, as permitted under the Copyright Act currently in force. Enquiries concerning reproduction outside these terms and in other countries should be sent to Oxford University Press at the address below

British Library Cataloguing in Publication Data
Data available

Library of Congress Cataloging-in-Publication Data
Van Beek, Steve, 1944–
  The Chao Phya: river in transition/Steve Van Beek.
    p. cm.
  Includes bibliographical references and index.
  ISBN 967 65 3069 7 (Boards):
  1. Chao Phraya River Region (Thailand) I. Title.
DS588.C414V36 1995
959.3'009693—dc20
                        94-26513
                         CIP

Typeset by Typeset Gallery Sdn. Bhd., Malaysia
Printed by Kyodo Printing Co. (S) Pte. Ltd., Singapore
Published by Oxford University Press,
19–25, Jalan Kuchai Lama, 58200 Kuala Lumpur, Malaysia

Preface

THIS book grew out of ruminations about the Chao Phya River while I sat on the veranda of a river-bank house opposite the Grand Palace in Bangkok. The stilt-borne wooden home I occupied for eleven years provided me an excellent vantage point to reflect on the myriad uses of the water flowing past my door. Curiosity about the river's origins led me to trek to the headwaters of the Ping and then to a village where I engaged a local carpenter to build a small boat.

The subsequent two-month journey to the sea answered some questions but ultimately posed even more, whetting my appetite to descend the other three tributaries, an odyssey that, in all, would occupy five months of my life. These solo journeys would not have been possible without the help of the hundreds of riverside villagers who took me into their homes and patiently explained what the river meant to them and how they employed it in their daily lives. It is to them that this book is dedicated.

Bangkok,   STEVE VAN BEEK
March 1994

Note

Throughout the text, unless otherwise stated, the term 'Chao Phya' refers to the entire river system—the Ping, Wang, Yom, Nan, and Chao Phya.

Acknowledgements

GRATEFUL thanks are extended to dozens of persons who made this book possible. Among the many who devoted special effort are:

Anek Chandarawongse
Scott Christensen
Andrew Clark
Harvey Demaine
Dhiravat na Pombejra
Mamta Giri
Roy Hudson
Craig Johnson
Bertil Lintner
Kathryn I. Matics
Nid Hinshiranan
Orawan Sriudom
Gideon Oron
Panapat Sahadrungsee
Phonchai Klinkhachorn
Phuthorn Bhumadhon
Jenny Piastunovich
Teddy Piastunovich
Plubpung Kongchana
Pornpimon Trichot
Sajee Charoenying
Srisakra Vallibhotama
Suphat Vongvisessomjai
Tanomwong Lamyodmakpol
Thaweesak Charitkuan
Theeraphap Lohitkul
W. J. Van Liere
Kirby Vining
Virat Khao-uppatum
Mandy Walsh
Yodying Maskasem
William Young

Contents

Preface — *v*
Acknowledgements — *vii*
Note on Transliteration — *xi*
Introduction — *xiii*

1 The Five Rivers and Their People — 1
  The Chao Phya River System — 1
  Naming a River — 5
  Geologic Origins — 7
  The River in Prehistory — 8
  The Five Channels — 11
  The First Tillers — 12
  The *Muang Fai* System — 14
  Central Plains Irrigation and Agriculture — 18

2 Liquid Road to Riches — 21
  Sukhothai's Dual Cultures — 24
  Ayutthaya's Rise — 29
  Looking Westward — 33
  River Engineering — 38
  Fortifications on the Lower River — 41
  Ayutthaya's Fall — 42

3 Link to the World — 44
  The River in War — 47
  The Return of the West — 49
  Floating Communities — 54
  Upriver Commerce after 1880 — 56
  The Rise of the Timber Industry — 58
  New Perspectives, 1900 — 60

4 Adapting Life to the River's Rhythms — 68
  House Design and Settlement Patterns — 68
  Rice Cultivation — 72
  Fishing — 74
  Boats — 79

5 The Spirit World and Cultural Life — 94
  Thai Attitudes towards Water — 94
  Water Festivals — 96

	Rituals	*100*
	Rites of Passage	*101*
	Buddhist Rites	*103*
	Village Celebrations	*104*
	Tribal Rites	*106*
	Perceptions of the River	*106*
	River Spirits	*108*
	Water-borne Processions	*110*
	The River and the Classical Arts	*114*
	The River in Folklore	*116*
6	**Taming the River**	*121*
	Portents of the Future	*122*
	A New Approach	*124*
	1902: Watershed Year	*125*
	New Directions	*129*
	Eroding *Muang Fai* Hegemony	*130*
	Major Lower-river Irrigation Schemes	*134*
	A River Rampant	*134*
	Water-borne Diseases	*136*
	New Perspectives	*136*
	The Effects of Dam and Irrigation Projects	*141*
	The Lower Portions of the Four Tributaries	*144*
7	**River under Siege**	*148*
	The Critical Shortfall in Water Supply	*148*
	Reasons for the Shortages	*150*
	Complicating Factors	*154*
	Environmental Problems in the Lower Basin	*162*
	Loss of the Power of Taboos	*169*
8	**Shaping a New Role for the River**	*172*
	Rethinking Objectives	*173*
	Reassessing Priorities	*177*
	Macro Responses	*181*
	Improving Water Use Efficiency	*182*
	New Technologies	*185*
	Returning Rivers and Canals to Their Former Roles	*187*
	New Attitudes Fostered	*188*
	Waste Disposal Techniques	*188*
	Innovative Agriculture	*189*
	Land and Forest Management	*190*
	Dams and Diversions	*192*
	Contemplating the Future	*197*
	Glossary	*199*
	Bibliography	*201*
	Index	*207*

Note on Transliteration

AFTER centuries of debate, there is still no common agreement on the proper transliteration of Thai words. The Thai Royal Institute system preserves the Sanskrit and Pali etymological spellings which are unhelpful to the non-specialist and unpronounceable except by Thai speakers. The International Phonetic Alphabet (IPA) widely used by scholars and lexicographers provides a tight phonetic rendering but is still unpronounceable by the non-linguist.

The system adopted in this book closely approximates the IPA rendering, but without the special linguistic symbols necessary for precise pronunciation. This sacrifice is in the interest of readers uncomfortable with the IPA but is at the expense of scholars who prefer the IPA's subtleties. Thai has contrasting vowel lengths and five distinct tones, none of which are specified here, also for the benefit of the non-linguist.

Certain widely known terms and names are spelt in their popular form rather than according to their phonetic system (i.e. Chao Phya, Loy Krathong, Khlong Toey, etc.). Direct quotes and bibliographical references retain their original spellings. Names of towns are spelt according to the Royal Institute, and geographic referents according to Royal Thai Army Survey Department maps.

After much deliberation, the author decided to render the plural nominative of 'Thai' as 'Thais' and 'Khmer' as 'Khmers' to avoid forcing the reader to decide if a word should be read as a noun or an adjective, or having to follow 'Thai' with the awkward 'peoples'.

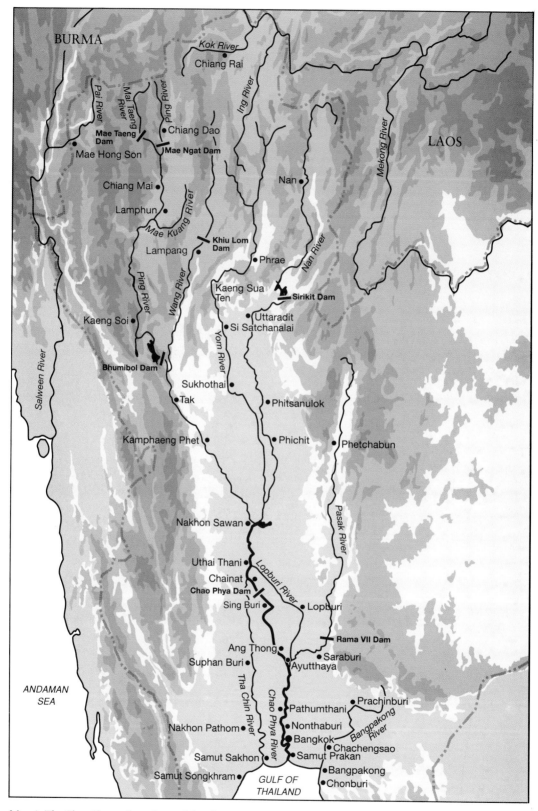

Map 1. The Chao Phya, tributaries, and distributaries.

Introduction

I do not know much about gods, but I think that the river is a strong brown god—sullen, untamed and intractable, patient to some degree, at first recognized as a frontier, untrustworthy, as a conveyor of commerce; then only a problem confronting the builder of bridges. The problem once solved, the brown god is almost forgotten by the dwellers in cities.[1]

FOUR thin blue veins descending from the upper edge of the page are the only indication the atlas provides of the grandeur of the Chao Phya River system. As they wriggle through craggy ranges, they are augmented by capillaries of *huai*, small streams flowing from side valleys which thicken the veins to a semblance of majesty.

At the southern periphery of the northern hills, the quartet descends into the broad alluvial Central Plains where two of the rivers flow into the other two. Still further south, the now-engorged twins collide, virtually at right angles, the brachiation conjoining at Nakhon Sawan. It is here, at the virtual heart of the Central Plains that the Chao Phya proper begins. As if sensing its ultimate destination, the meandering river takes a bolder course, its bends flattening and broadening as it moves southward. Blue ink flows down the atlas page counter to compass point through brown and green before entering the country's vibrant heart, Bangkok. A few dozen kilometres below the kingdom's capital, the line splays to become a mass of blue inscribed 'Gulf of Thailand'.

The atlas's static blue line provides little indication of the river's character or of the vital role the Chao Phya has played, and continues to play, in Thai consciousness. It is not a raging torrent like many great rivers of the world; its shallow gradient mitigates against whitewater rapids of thunderous ferocity. But neither is it a passive force, its waters merely bearing Thailand's history to the sea. As the principal shaper of Thai heritage, culture, and economics, the river has moulded the nation's values, been a life source for its rich agriculture, served as a highway for commerce, linked its cities, and barred its enemies.

Return to the top of the atlas page. A finger sliding down the four blue veins—the Ping, Wang, Yom, and Nan—traces the political and economic development of Thailand from its thirteenth-century inception as a nation to the present day. If one accepts the theory that Thailand was populated by Tai *émigré* filtering south from China, it was to the northern river valleys that they migrated. If one favours the thesis of proto-Thais emerging from caves to claim their destiny, it is equally evident that they chose to carve their villages and farms from the thick jungles crowding the fertile river-banks.

Thailand's history as a nation begins in the north. Over the centuries, kingdom succeeded kingdom, moving steadily southward along the rivers: the Sukhothai kingdom (1238–1350) near the Yom River, Ayutthaya (1350–1767) on the Chao Phya, Thonburi (1767–82) on the right bank of the Chao Phya, and Bangkok (1782–present) on the left bank. Most of Thailand's major central and northern minor kingdoms were also founded on the banks of the Chao Phya or its affluents.

The moving finger also traces Thailand's economic development. It begins in the North with slash-and-burn hilltribe agriculture, passes through the rice fields of the Central Plains, and ends in the industrialized cityscape of Bangkok. While other areas of Thailand have remained comparatively poor, the northern and central regions etched by the Chao Phya have become economic powerhouses and major population centres.

Some scholars contend that Sayam, the name by which Thailand was known until this century, translates as 'people of the river',[2] the 'river' denoted being the Chao Phya. Rivers and water symbolism are inextricably bound up with Thai village and cultural life. Water is the incubator for the staple foods of rice and fish. It is the principal element in rites of passage and in Thailand's two most important festivals, Songkran and Loy Krathong. Its soft sensuality pervades the flowing lines of Sukhothai Buddha images, and the sinuous lines and planes that proliferate in *wat* (monastery) architecture and mural paintings. The river provides an allegory of the Thai mode of negotiating life's obstacles: it does not confront them, it flows around them. Thais do not live as independent entities, they blend their lives together, melding through consensus and compromise to preserve a liquid continuity whose surface, while often masking turmoil and contradiction, serves to lubricate social interaction.

For centuries, the rivers coursed untrammelled, with the movements of travellers using them as roadsteads dictated by the waters' whims. From the 1700s, boatmen poled wooden barges up the Ping on six-week journeys to Chiang Mai. They timed their return voyages to coincide with the monsoon flows, braving the rapids above Sam Ngao, and then paddling the flatter stretches of the lower river to the sea.

The introduction of steam-driven launches in the mid-1800s changed travellers' perceptions of the river. Rather than flowing with it, they subverted the Chao Phya's force to the brawn of powerful engines, using the waterways as mere motive surfaces on which to glide to distant destinations. By the 1970s, the river had all but ceased to serve even that purpose. Highways were built far from the rivers, and commerce streamed along them. Far-sighted riparian entrepreneurs relocated their businesses to the towns that began springing up on highway shoulders, consigning the river-banks to history.

In the 1950s came the dam builders who treated the river, not as a natural entity, but as a component in an evolutionary progression towards prosperity. Over the next twenty years, engineers would turn portions of the Chao Phya and three of its four tributaries into still ponds to light cities, power industry, irrigate crops, and quell floods. The intricate traditional irrigation systems of the North would be abandoned for diesel pumps which sucked water as if from a storage tank. The rivers would become mere ditches to guide water to other destinations, with side canals diverting the silent waters from their journey through the mountains to the sea.

Today, the Chao Phya River is seriously endangered, a provider whose very existence is threatened by the people it serves. Its original importance forgotten, it is regarded as an exploitable adjunct to development, transformed from a vital ecosystem into a conduit for irrigation and drinking water, and for waste. Contrary to popular belief that its degeneration is the work of lower-river urban and corporate interests, it is being destroyed along its entire length, from headwaters to mouth. Towns of all sizes deposit trash on its banks, factories flush sewage and chemicals into its flowing waters. Huge dredges cut deep into banks and bottoms, scooping up sand for construction projects. The process alters the river's course, disrupts its hydrology, kills its fish, and erodes farmlands. Ships moored in the harbour discharge oils which destroy the mangroves and poison the oyster beds in the Gulf.

In the villages, trees are cut, allowing the growth of brambles which choke waterways and invade farmlands. Without tree roots to cement the soil, river-banks crumble and flow away. Heedless of the consequences, fishermen employ destructive techniques to catch fish. Pesticides and herbicides seep into the water during monsoon rains, and detergents are washed into it, bringing eutrophication and death to large numbers of fish. Like the houses that now face away from the rivers, Thais seem to have turned their backs on the Chao Phya.

And yet the myth of its greatness persists. Thais refer with pride to the Chao Phya's ancient lineage as the River of Kings. It is an important psychological conceit which anchors their faith in nationhood, and buoys them in their progress towards material prosperity.

This book attempts to document the changing beliefs about, and uses of, the Chao Phya River system over the centuries. Its purpose is to discover how an entity which performs a vital role in daily commerce and life is being destroyed by seemingly contradictory aims and attitudes. At times this book may seem to be a polemic on short-sightedness. It is meant, however, not as a crucifixion but as a hope for resurrection.

INTRODUCTION

1. T. S. Eliot, 'The Dry Salvages', in *Four Quartets: The Complete Poems and Plays of T. S. Eliot*, London: Faber and Faber, 1969, p. 184.
2. Chit Phumisak, *Khwam Pen Ma Khong Kham Sayam, Thai, Lao le Khawm, Le Laksana Thang Sankhom Khong Chu Chon Chat* [The Origins of the Words Sayam, Thai, Lao, and Khom and Social Characteristics of These People], 2nd edn., Bangkok: Samnak Phim Duang Kamol, 1981.

1 The Five Rivers and Their People

The influx of Thai-speaking peoples from the North [after AD 200] was a drawn-out infiltration spanning centuries.... Thai–Lao speakers descended the rivercourses and settled along the tributaries of the main rivers on the riverbanks, or in or along the floodlands.[1]

The Chao Phya River System

THE pawn of gravity, geography, and geology, the etcher of an ever-changing path across the landscape, the Chao Phya is Thailand's single most important river system. Draining approximately 117 500 square kilometres, within a region representing 35 per cent of the country's total land area, it nourishes rich alluvial farmland that supports nearly one-half of the nation's population. Excluding canals and minor tributaries, an annual average of 16 700 million cubic metres of water flow from the northern border to the sea along a 2925-kilometre network. One of Asia's most important river systems, it comprises four tributaries and a main channel ordered, from west to east, by the litany that Thai schoolchildren have intoned for ages:

Maenam Chao Phya mi si sai:
Ping, Wang, Yom, Nan.
Maenam Chao Phya lai pai Paknampo.

The Chao Phya has four tributaries:
Ping, Wang, Yom, Nan.
The Chao Phya flows from Paknampo.

The westernmost of the four, the Ping rises in the Dan Lao Mountain Range (Plates 1–2). It trickles down the slopes of an unnamed 1824-metre-tall mountain, 8 kilometres north of Ban Doi Tuai, a Musur Daeng hilltribe village on the Thai–Burmese border, 115 kilometres north of Thailand's chief northern city, Chiang Mai. On a 715-kilometre journey south from its headwaters, it 'stairsteps' down a series of passes and plains before spilling through a long gorge to empty into the Chiang Mai Valley. Exiting the valley, it skirts the craggy Tenasserim Range that defines the border with Burma. Passing Chom Thong and Hot, it enters the 120-kilometre-long Bhumibol Dam reservoir which has silenced what were formerly the most treacherous rapids of any Thai river.

1. At the headwaters of the Ping River on the Burmese border, a Musur tribesman propitiates the spirits of the river in what is said to be the footprint of a mythical elephant.

2. The upper Ping River is still cloaked in thick forest.

THE FIVE RIVERS AND THEIR PEOPLE

Below the Bhumibol Dam, the Ping turns south-east through the old logging town of Raheng (Tak) where it shallows and broadens from 80 to 450 metres, the widest point of its length, and the widest of any tributary. Narrowing to 100 metres, it passes the historic town of Kamphaeng Phet, and finally enters the city of Nakhon Sawan, formerly known as Paknampo (and still referred to as such by rural residents) meaning 'the mouth where the waters co-mingle'. Here, the Ping's clear green waters blend with the brown waters of the Nan River to become the Chao Phya.

At 400 kilometres, the Wang River is the shortest of the four tributaries. It originates in the Phi Pan Nam Khun Wang Haeng (Mountain of the Spirits Who Give Water) on the flanks of the 1453-metre-tall Doi Maeo Ta Mao in the Doi Luang National Forest, 80 kilometres south of Chiang Rai. Slithering south through the mountains, it is slowed momentarily by the Khin Lom Dam before reaching Lampang, another historic town made important by a river. Below Lampang, the Wang flows through farmlands to join the Ping River just downstream from the Bhumibol Dam. The combined Ping and Wang are often referred to as the Khwae Kamphaeng (Kamphaeng Branch), the Khwae Noi (Small Branch), and the Ping Thae (True Ping).

Twinned like the Ping and the Wang, the eastern tines of the fork, the Yom and Nan, are known as the Khwae Yai (Large Branch). The as yet undammed Yom is generally considered to flow from its principal tributary, the Nam Kha, whose own tributary, the Huai Nam Yip, rises at Doi Thiu Khao Dan Lao in the Doi Phu Kha Range, 75 kilometres north-east of Phayao, and 70 kilometres north-west of Nan town. The Nam Kha slices south through rugged mountains for 40 kilometres to join the Nam Khuan at Amphoe Pong to become the Yom. Shortly thereafter, it enters a long gorge, where it picks up speed to hurtle through the Kaeng Sua Ten (Dancing Tiger Rapids) to Phrae. An east–west mountain range then forces the Yom north-west to Amphoe Wang Chin where, rounding the end of the range, it flows south again. It then enters the crucible of Thai history, the Sukhothai Valley, passing first the ancient town of Si Satchanalai, and then the modern town of Sukhothai, 12 kilometres east of old Sukhothai, Thailand's first capital city. Paralleling the Nan River, at times passing within a few kilometres of it, the Yom completes its 700-kilometre peregrination at the *chedi* of Wat Ban Kluai Chai, where it enters the Nan about 30 kilometres north of Nakhon Sawan.

Thailand's longest river, the 740-kilometre Nan begins near the Luwa hilltribe village of Baw Khlua on the summit of the 1300-metre-tall Doi Kun Nam, in the Luang Prabang Range that forms the border with Laos. The Nan rushes north-west through deep forests (Colour Plate 1) for 30 kilometres before turning south to enter the town of Nan. Leaving the lower end of the fertile Nan Valley, it carves a narrow path through a long, rocky gorge. Exhausted at the end of its long run, it is absorbed by the reservoir of the Sirikit Dam. Leaving the dam spillway, it flows south-east into Uttaradit and then south through Phitsanulok. Past Phichit and Chumsaeng, the Nan is joined by the Yom before reaching Nakhon Sawan.

For more than a kilometre below Nakhon Sawan's riverside market, the brown waters of the Nan decline to mix with the green waters of the Ping. Instead, they flow side by side, separated as if by invisible walls, until finally they meld into a single chocolate band. The amalgam of four rivers, now known as the Chao Phya, begins a 370-kilometre surge towards the sea. From Nakhon Sawan the broad, muddy river descends from an elevation of 23.5 metres above sea-level, spilling over the lip of land and into the delta at Chainat. As it flows past Ayutthaya, it is joined by the 570-kilometre Pasak, sometimes regarded as a fifth tributary. At Ayutthaya, the river stands only 3.5 metres above sea-level and is nearly at sea-level when it cleaves Bangkok and Thonburi into twin halves. The river completes its journey 7 kilometres below Samut Prakan (Paknam), sliding across a high mud-bank to enter the sea.

Having become the conjoining of four rivers at Nakhon Sawan, the Chao Phya then gives birth to three children of its own, lending it the semblance of a tree whose lower trunk divides into sturdy roots anchored in the watery pot of the Bight of Bangkok, the northern extremity of the Gulf of Thailand. The first river, the Tha Chin, also called the Suphan Buri and the Makham Thao, branches west from the Chao Phya 77 kilometres south of Nakhon Sawan, or 7 kilometres above Chainat. Running nearly parallel to the Chao Phya, it passes Suphan Buri and enters the Gulf at Samut Sakhon. The second river leaves the Chao Phya 5 kilometres south of Chainat. Often referred to as the Menam Noi or Chao Phya Noi (Lesser Chao Phya), it is one of the former courses of the Chao Phya itself, a channel that flows through Amphoe Sena before rejoining the Chao Phya River at Bang Sai.

Although it is termed a river, the Lopburi is really a loop of the Chao Phya. It flows due east from Sing Buri, then winds south to pass the town of Lopburi and enter the Pasak River. Ultimately, via Khlong Bang Phra Khru, it joins the Chao Phya at Ayutthaya. In addition to the rivers, there are the thousands of kilometres of natural and excavated canals, capillaries to the great arterials. From the air, these appear as one great bedewed spider's web stretching to the horizon, a complex system of waterways that is testament to nature's bounty and to human tenacity in etching paths across the landscape.

While the gradient near the mouth is low, the river-bed itself lies quite deep below the surface. Measuring it in 1828, John Crawfurd found it to be 'not less than nine fathoms' (16.4 metres) at Bangkok and seven fathoms (12.8 metres) at Paknam.[2] Before it was dredged in the twentieth century, the river-mouth was blocked by mud-banks, barring entry to ships with draughts greater than 2.5 metres until the high tide lifted them gently over the earthen gateway to the kingdom.

The lower Chao Phya's level and volume is regulated both by seasonal variations in rainfall and by tidal flux which reverses the river's course twice each day. With the onset of the monsoon rains in May and June, the Chao Phya begins to rise but much of its water spills laterally into the parched rice fields. By July, after the early storms have blown themselves out, the rainfall tapers off. After the year's heaviest downpours in September, the river flows in full torrent. The flat delta ensures that the lower Chao Phya is tide-affected from Ang Thong to the sea, a distance of 180 river kilometres. The downriver monsoonal flow is met by a bi-monthly peak tide pushing upstream from the Gulf. Compressed between these two forces, the river brims its banks, normally flooding the delta south of Pathumthani once every two weeks: in mid-September, twice in October, and once in November. From mid-November, the waters abate and the river drops well below its banks. In April, water-levels in the Ping and Yom around Tak and Sukhothai fall so low that it is possible to walk across the rivers with barely a wetting.

Naming a River

The generic terms for rivers describe their sizes. *Huai* and *lam* are northern Thai words for 'stream'. *Nam*, strictly translated, means 'water' but refers to a small river, while *nam mae* indicates a slightly larger river, and *maenam* or 'Mother of Waters' is reserved for major rivers. *Khlong*, or 'canal', is derived from a Mon word meaning 'path' and refers either to a natural or to a man-made waterway connected to a river.[3]

It is not known by what name the Chao Phya was originally called or how it acquired its Pali-based regal honorific, the title of a high-ranking court official. The name first appeared in a 1687 Franco-Thai treaty during the reign of King Narai (1657–88), but the reference was to a town, Bang Chao Phya, at the mouth of the river. In a letter to HRH Prince Narisara Nuwattiwong, historian HRH Prince Damrong Rajanubhab noted that a BE 2295 (1752) message from the King to a Thai ambassador escorting a delegation of Thai monks travelling to Sri Lanka, contained the comment that 'the populace accompanied the monks to the "Pak Maenam Bang Chao Phraya"', apparently a reference to Samut Prakan.[4] Some writers have suggested that the river's name derives from that of the village. Since, however, the river-mouth was formerly located at Phrapadaeng (and only when the delta had advanced into the sea was the village founded) the village would have taken its name from the river, not the reverse. Historically, Thai monarchs have been involved in the control and provision of irrigation water, so it may be logical to assume that *Chao Phya*

THE CHAO PHYA

alluded to the king's role as lord of the waters.

The river appears to have been known to most Thais simply as 'Maenam' (the River), which described the entire system with its many tributaries and distributaries. Nicolas Gervaise, a French cleric resident in seventeenth-century Ayutthaya, wrote,

> Nor will I record the names of the rivers I have mentioned above, because the Siamese only call them after the large cities through which they flow. The [foreign] authors of earlier accounts of the kingdom call the great river of Siam the *meenam*, but this is because they are ignorant of the language of the country, for in Siamese *meenam* simply means a river.[5]

While correct in his assertion, he, Simon de la Loubère, and all other writers of the period continually refer to the Chao Phya as 'the River' or 'Menam'. The Ayutthayan Chronicles are silent on the subject.

By 1828, it was still called 'Menam' by diplomat John Crawfurd, who drew on respondents within the Thai court for his information. He notes that:

> The word Menam literally means 'mother of waters', and seems the only one in the Siamese language for a river. It is in fact a generic and not a proper name; and although applied to the great river of Siam (*par excellence*, the river) it is equally applicable to any other. In the Siamese

3. The Chao Phya slices through the heart of Bangkok and its twin city of Thonburi.

language, as in many other Asiatic ones, there seems to be no one proper name to distinguish a river throughout its course, each separate portion of it taking its name from the principal place by which it passes, as the river of Bangkok, the river of Kampeng-pet, the river of Chang-mai, &c. which mean only different parts of the Menam.[6]

Three decades later, Rama IV (King Mongkut) was bemoaning British ignorance about Thai geography, writing that 'according to them ... the river running through Bangkok [Plate 3] has no other name but "Menam"',[7] although it is apparent that the Thais themselves still referred to it by the same term. Even as late as 1904, an official handbook, written to accompany a Thai exhibition to the Louisiana Purchase Centenary in the United States, noted that the river was commonly referred to as the 'Menam'.

Geologic Origins

The river's story begins long before it had a name or even an identity. Tracing its evolution and its myriad personalities requires reaching far back in time to the period when it was still seeking a path and a purpose: '... when the world was not yet in existence, there ... was nothing except the air ... cold and ... hot [which] gave rise to a wind that blew very violently and continually. The cold, in contact with the earth, produced dew, mist, fog and rain which started falling.'[8]

Written between the thirteenth and sixteenth centuries, the Thai Yuan[9] manuscript, the *Pathamamulamuli*, relates the genesis of the earth and its rivers. While its language is more poetic than modern scientific texts, it is evident that its author understood well the cycles of evaporation, condensation, and rain which create and consume rivers. By contrast, the geologist's explanation for the river system which has defined human life in the North and Central Plains is more prosaic, but no less enthralling.

The ebb and flow of aeons of global warming and cooling gave rise to a plethora of Thai rivers, especially in the lower Chao Phya basin. Each time the land was exposed, a new set of rivulets began to flow, ever-changing tendrils that bore little resemblance to the rivers known today.

The upper river basin was shaped 250 million years ago during the Permian era by compression from the west which folded sedimentary rocks along a north–south line and extruded the lower granitic and other igneous basal rocks. The higher elevations occurred along an imaginary line running east and west approximately the latitude of Phayao. Gravity impelled rivers like the Kok and Ing to flow north into the Mekong and the four tributaries to flow south. The steepness of the gradient and the density of the rock dictated that the northern portions of the tributaries changed course only moderately over the aeons. Even in the northern valleys created by the sediments they carried, the rivers meander only slightly, suggesting they are still quite young.

Block faulting created the graben depression that characterizes the middle basin extending from Phitsanulok down to Nakhon Sawan where, until well into modern geological history, the rivers seem to have emptied into the sea. Over the centuries, eroded soil consumed in the mountains was expelled by the river-mouth to build an alluvial fan that pushed the Gulf of Thailand further south.

The climate during the late Pleistocene (50,000–20,000 BP) was probably cooler and drier than at present and the land was likely divided into only two environmental zones: a northern dry seasonal parkland and a southern parkland or savannah. Human habitation seems to have been concentrated along rivers and near forests.

The Pleistocene Ice Ages lowered sea-levels, exposing the Great Sunda Shelf, an enormous land mass that extended south to Bali and east to the Philippines. When global warming melted the ice-caps during the late Pleistocene between 14,000 and 45,000 years ago, the rising seas inundated Thailand as far north as Uthai Thani, re-creating the great bay that had existed aeons before.[10]

By the Early Holocene (8,000–12,000 years ago), the climate was similar to that found today and the land was covered in great stands of rain forest. The river valley was inhabited by people described as Hoabinhian after the Hoa Binh archaeological site in northern Vietnam. As anthropologist Douglas Anderson notes, 'The proximity of the Hoabinhian sites to the rivers suggests that the rivers were used not only as a source of food but also for transportation. In view of the long history of the use of water-craft by Asians, the use of boats by Hoabinhians is not at all surprising.'[11]

The Quaternary period saw the laying down of alluvial sediments over the granite block base and the creation of marine clays that would constitute the bed of the lower river and the rich land running either side of it. During the Middle Holocene, settlements dotted the coastline and both sides of the rivers, extending laterally into the foothills of the mountain ranges. Temperatures 2–3 °C higher than at present melted polar ice-caps and caused the sea to rise again. At its highest point in the Holocene Optimum, 6,500–7,300 years ago, the sea covered most of the present Chao Phya basin, a fact which has been established by carbon-14 dating.[12] As sea-levels declined, the land was exposed yet again and the prototypes of modern Thai rivers began to flow, cutting channels and creating the geography that characterizes Thailand today.

The River in Prehistory

To describe the modern Chao Phya of the past 5,000 years as treading a single, unwavering path is to misrepresent its evolution. The lower Chao Phya is a young river that in its infancy acted like an old river, wandering aimlessly with no seeming objective. With few mountain barriers to levee it, and a bed in alluvial soils with a

THE FIVE RIVERS AND THEIR PEOPLE

virtually flat gradient, it is little wonder that it meandered, trying one path for a while then, blocked momentarily, seeping in a new direction until it established a new channel. To postulate the sequence in which the courses were created is difficult without more sophisticated dating methods. In a soft bed which leaves no marks, only creases and sediments, a river's wayward course provides few footprints to trace its chronological development. Aerial views reveal hundreds of now-dry loops that suggest the river's continual change of mind. It is likely that it had several simultaneous courses as it does now, the Tha Chin, the Lopburi, and the Maenam Noi Rivers being little more than branchings from the Chao Phya itself.

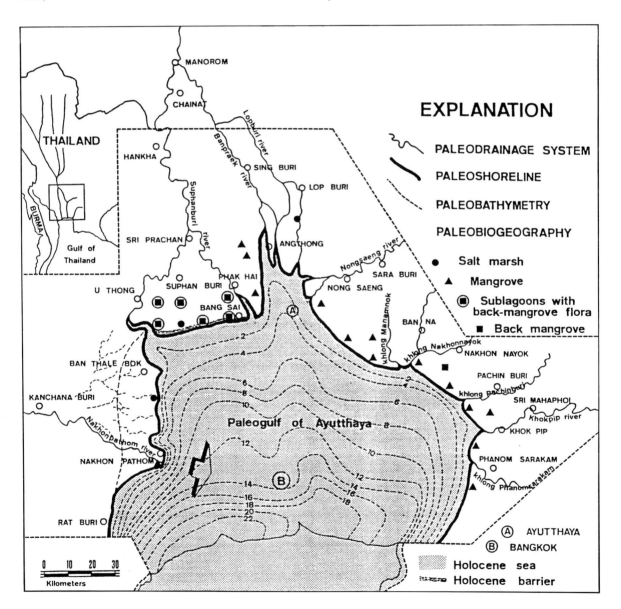

Map 2. The Holocene embayment of the lower river with proto-rivers Suphan Buri and Bangpraek. (Courtesy of Pergamon Press)

It is known that between 3,000 and 6,000 years ago, the Gulf of Thailand extended far north into the valley, with Ratchaburi, Nakhon Pathom, Ayutthaya, Nakhon Nayok, and Prachin Buri as coastal towns ranged along the edges of a broad bay (Map 2). The chief waterways were two rivers which flowed south from Nakhon Sawan and forked at Manoram. The larger, western branch, the Suphan Buri River, continued south and entered the sea at Bang Sai, 20 kilometres west of present-day Ayutthaya. The eastern branch, the Banpraek, flowed south-east to enter the Gulf just north of present-day Ang Thong.

As the seas began to recede from the basin about 3,000–4,000 years ago, a new stream, the proto-Chao Phya, breached a channel in the sediment wall, cutting its way down to Chainat and from there to Ayutthaya. With time, it came to carry the larger volume of water and the Suphan Buri was reduced in importance. The Banpraek shrank to a fraction of its previous size and today is little more than a broad canal.[13] As the water-levels fell and the former Gulf bottom was exposed as arable land, settlements begin to appear on the shoreline. Once the rivers had stabilized, people built homes on their banks. Chinese annals attest to the existence of cities in the basin from the seventh century AD. Most were in the area of Nakhon Pathom near the Tha Chin River and, to the east, in the area of the Bangpakong River. The marsh was broad enough to split the basin's population into two distinct cultures. Those at Nakhon Pathom, Kubua, and U-Thong, west of the river, practised Buddhism, while those at Si Mahasot in the region of the Bangpakong River, east of the marsh, professed Hinduism.[14]

Most interesting were the primitive settlements at U-Thong, which may have flourished as early as AD 100. Aerial photographs of U-Thong suggest that even at this date, valley dwellers may already have been demonstrating skills as canal engineers. The photographs show a thick, 13-kilometre straight line running east from U-Thong to what geologic studies indicate would have been the head of the Gulf. Given the placement, the line can be nothing other than a canal, a surprising development since it indicates not only a considerable degree of engineering skill and social organization but sufficient manpower to dig a large navigation channel. The channel likely resembled the narrow canals one sees when flying along the Chonburi coast, channels which run from inland villages through the mud-flats, enabling fishermen to reach the sea at low tide.

About 10 kilometres or three-quarters of the way down the main canal from U-Thong, another canal branches to the south-east for about 5 kilometres. It appears to have been dug at a time when the combination of falling sea-levels and river-borne silt filling the delta was forcing the Gulf to retreat. As the Gulf's northern shoreline advanced southwards, the original canal was no longer functional. The new, diagonal canal would have ended at the northern boundary of the newly shrunken Gulf, a few kilometres south of its former position. For the same reason, a third canal, dug at a later date, branches to the south-east, 6 kilometres from U-Thong, and runs parallel to the first branch. It ends some 8 kilometres below the first branch and 11 kilometres south of the original canal, indicating that the sea had retreated some 11 kilometres from its maximum northern extension. As the Gulf continued to shrink, the canals were abandoned and U-Thong became land-locked. It is difficult to establish a precise chronology for the canals but, given the shallowness of the Gulf to begin with, it would have taken less than a half-metre decline in sea-level to reduce the Gulf by 11 kilometres in a very short period, perhaps less than 100 years.

The Five Channels

With such a shallow gradient, it is not difficult to imagine a river's eating away at an embankment during several flood seasons and finally breaking through to establish a new course. What is surprising is that the Chao Phya may not have achieved its present course until the mid-nineteenth century.

No precise dating has been established for each course shift but a rudimentary timeframe has been developed by Thaweesak Jaritkhuan in his study of the river.[15] He places the first of a series of five principal channels in the early Dvaravati period (sixth–eleventh centuries AD) after the river-mouth had moved 100 or more kilometres south of U-Thong. As indicated by Map 3, the original Chao Phya (now referred to as the Chao Phya Noi) flowed west of its present course beginning just below Chainat and running to Ang Thong via Sankha Buri and Phrom Buri. After some time, it appears to have silted up and it shifted to the east to nearly its present channel via In Buri and Sing Buri.

The river must then have breached a wall in its right bank because in its route it ran south-west from Ban In-Pramun (15 kilometres north of Ang Thong on the Chainat–Ang Thong canal), to Bang Sai via Wiset Chai Chan, Phak Hai, and Sena, and then entered its present-day channel. In its third incarnation, it broke its banks at Chainat, flowing nearly north-east for 12 kilometres before turning south, approximating the Chao Phya's present course from Chainat to Bang Sai. Instead of passing Ayutthaya as it now does, however, it branched at Wat Chulamani, 12 kilometres north of Ayutthaya, and flowed south-west, entering Chao Phya Noi 8 kilometres east of Sena and then winding past Bang Sai and heading south.

The fourth route, dated to 1813 in the reign of King Rama II (1807–24), is the work of engineers. The objective was to straighten the Chao Phya to flow through Ayutthaya. Workmen dammed the Chao Phya at Ang Thong, causing it to proceed east and then south along the shallow Khlong Bang Kaew. In the vicinity of Ban Mai (8 kilometres north-west of Ayutthaya), it would have flowed

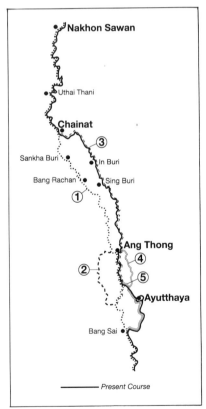

Map 3. The five most recent channels of the Chao Phya before it assumed its present course.

south through Maha Rat where it poured through a 2-kilometre canal into the Lopburi River. It re-entered the old Chao Phya River at Wat Mai, about 5 kilometres south of Wat Chulamani. The engineers hoped that the flowing waters would erode and scour the canal, deepening it to serve as the main river. The attempt failed because water prefers to flow in a straight line and the man-made barrier impeded it. The dam broke repeatedly under the relentless pressure of the annual floods and the project was ultimately abandoned.

It was not until 1857 that an alternative path was created. A 5-kilometre channel was dug from the entrance of Wat Chulamani to Ban Mai. The river responded by following this new course and abandoning the old one, in effect making a secondary river of the stretch that ran from Ban Mai, and into the Chao Phya Noi. Half as wide as the river above and below it, the 1857 Ban Mai shunt funnels the Chao Phya down to Ayutthaya.

While the Thai kingdom was being established at Sukhothai in the thirteenth century, alluvial deposits were pushing the delta and river-mouth further into the Gulf. It is on record that around AD 1050 the Chao Phya's mouth was at Phrapradaeng, 23 kilometres upriver from its present location. The Khmers built a town there that served as a seaport until King Songtham (r. 1610–28) founded Samut Prakan (Paknam).

It is evident from the distance between present-day Samut Prakan and the sea, that the Chao Phya has continued to move sediment southwards into the Gulf. Comparisons of river charts drawn in 1797 and 1856 suggest that the mouth moved 4 kilometres in 60 years, an average rate of 67 metres per year. The rate of expansion after that slowed. In the twentieth century, the eastern coastline has grown marginally while the western shore has been moving east at 5 metres per year, creating a lip so that the river now curves south-east before entering the Gulf.

It is clear that the river's history has been one of responses to changing geological and climatological seasons. An aeon-long stop-frame movie of Thailand would reveal it as a gigantic clay-mation surface being kneaded into a variety of physiognomies as it settled into a semblance of the topography recognizable today. Having considered its geography, it is necessary to consider how the latecomers on the scene—humans—utilized it.

The First Tillers

Nourished by the five rivers, the fertility of the northern valleys and central plains drew waves of migrants who settled along or near the forested river-banks. Who were these people, what did the land and rivers offer them, and how did they manipulate these two elements to serve their needs? An examination of the early peoples begins in the north, the only area where material evidence of prehistoric cultivation has been found.

THE FIVE RIVERS AND THEIR PEOPLE

Wilhelm G. Solheim, Chester Gorman, and others have documented the existence in the northern hills of cave-dwelling peoples dating from 7600 to 12,000 BP. Strewn on the cave floors were plant remains similar to those of crops cultivated today, suggesting that even at this early date, these hunters and gatherers were also farmers. It is probable that they fished the streams and drew water from them for domestic purposes.[16]

Although small clusters of people thrived in isolated pockets, resources before 6000 BP were insufficient to support large, permanent settlements. Instead, groups of widely scattered foragers survived by 'hunting, gathering and simple gardening.... Fish and shellfish ... were not especially abundant.'[17] This changed as Holocene climate patterns stabilized. Now, Thailand and the rest of South-East Asia could begin cultivating a wide array of grain and legume crops using the draught animals they had domesticated. Two plants in particular—rice and taro—flourished in the marshes of the lower basin, laying the foundation for the area's economy.

The design of the first settlements established along the northern rivers from the sixth century AD onwards shows strong Chinese influence, suggesting the migration of a people who had either entered virgin territory or melded with an indigenous race. Northern Thai stilted houses, modes of rice cultivation, bronze drums, male tattooing, and cults of the *naga*, the potent water symbols that proliferate in Thai culture, have cultural antecedents in the Yue civilization of southern China during the sixth through the second century BC.[18]

How and when these Tai people named the northern rivers is a matter of conjecture. As with the term 'Chao Phya', the search for the origins of river names—none of which has a meaning—is an intriguing exercise in deduction. European writers of a later period, Nicolas Gervaise among them, suggested that rivers were named for towns on their banks, a contention that has persisted among several present-day writers. It is equally valid to suggest that semi-permanent riverine residents would have no need to give their river, or even their village, anything other than a generic name. Only when they encountered a second river and needed to distinguish between the two would names be given to each stream. Moreover, only when a village had grown to sufficient size to warrant a name would it acquire one. In the Shan States and in southern China, whose Tai-speaking people share common heritage with northern Thais, it is normal for a town and a river to carry the same name. But this does not resolve the chicken–egg dilemma about the origins of the four tributaries' names, especially since, unless the town names have changed, the only tributary of the four flowing past a town of the same name is the Nan.

A clue lies elsewhere. Linguistic and historical studies note that eleventh–twelfth century Angkorian inscriptions referred to groups of Thai mercenaries serving in Khmer armies as people coming from a particular river in Thailand. Moreover, the annals suggest that this is the way the mercenaries referred to themselves.[19] In the absence of firm evidence, it is tempting to postulate that the first inhabitants of the northern regions brought the names of the rivers from elsewhere. Thus, like those European immigrants to the New World who named their new towns New Amsterdam or New London, these Tai people, known as a people from a particular river, may have applied that name (or had it applied to them by others) to the new rivers. A look at a map of the Shan States of north-eastern Burma reveals the presence of both *nam* (stream) and *mong* (*mong* is the Shan equivalent of the Thai word *muang*, or town). A closer examination reveals the presence of Nam Ping and Nam Yom as well as their tributaries of Nam Teng (Nam Taeng in Thai), Hka (Kha), Yi, and Ing. Moving north-east into the Tai-speaking areas of China's Yunnan province, one encounters towns designated as *meng*. Notable among them are Meng Pin and Meng Peng, as well as Meng Wang, all on rivers of the same name. The duplication of names may suggest that the original name in Yunnan was carried by migrants to the Shan States and from there to northern Thailand. It may also suggest that two sets of emigrants left China, one bound for the Shan States and the other for the high valleys of northern Thailand. The lone exception is the Nan, for which no other reference has been found.

The Muang Fai System

Then, as now, northern farmers cultivated a round, glutinous strain of rice similar to that grown in Yunnan, the Shan States, and Laos. To ensure a steady supply of water throughout the growing season, water had to be transferred from the streams into the fields. It appears that twelfth-century northern immigrants brought with them such a system called *muang fai* which had been utilized in the kingdom of Muang Suwankhomkham (north of the Mekong River). The Lanna Chronicles suggest that these may already have been in existence in China for seven–ten centuries. The Yue peoples of Sipsongpanna are also known to have utilized a *muang fai* system to irrigate their crops.[20]

Still operational today on most northern rivers, the basic *muang fai* system appears to differ little from its ten-century-old predecessors. Designed for streams less than 20 metres wide, the system comprised weirs built in fast-flowing rivers to impound water for irrigation purposes. Most weirs were made of stout wooden stakes driven into the river-bed, with large stones lodged between them (Colour Plate 2; Plate 4). Few exceeded 2.5 metres in height, never high enough to entirely impede the flow of the river which brimmed the weir's lip and rushed on for use by the next *muang fai* system downstream (Plate 5). Instead, the dams raised the water to sufficient height to flow through a system of channels into adjacent fields. Regulator boxes at the heads of each channel

4. This *muang fai* weir is constructed primarily of bamboo stakes driven at an angle to the river's flow with stones piled behind it to impede the water's passage.

5. A more elaborate weir which allows excess water to spill over the lip for use by downstream farmers.

controlled the amount of water each field received. In this form, the *muang fai* system has parallels in such widely dispersed areas as Peru and Sri Lanka.

Whereas systems in remote areas were local initiatives overseen by local administrators, the operation of those in the more-populated regions were governed by northern monarchs. The Yonok Historical Records state that 'around B.E. 1300 [AD 757], King Choeng who had ruled [Chiangsaen] for 11 years ... ordered a big irrigation canal to be dug from Mae Sai River to bring water to the rice field on the left side areas to facilitate the people's rice planting'.[21] As the statement refers to a period predating by five centuries the recorded founding of the northern kingdoms, the date of AD 757 is likely to be apocryphal. Even allowing for the dating discrepancy, it still suggests the antiquity of the system.

Royal involvement can also be seen in the earliest of the Muang Fai laws which were incorporated into the Mengraisat laws issued by King Mengrai (r. 1259–1317) in 1296. They detail the tenets of the system and their impact on social organization. These contracts and records were passed from generation to generation and modified according to specific needs and changing conditions. As an indication of their comprehensiveness, the oldest version, found at Wat Saohai in Saraburi province, covers both sides of a 48-sheet palm leaf manuscript.

In the larger systems, the *khun nai fai* (irrigation official), later designated a *hua na fai* (irrigation headman), was appointed by the king and held office for life. He was aided by an assistant and, as the centuries passed and the system became more extensive, by a team of assistants with specialized tasks. Once the system was established, the position of *hua na fai* became an elected office, with the holder selected by a simple majority vote by his fellow villagers.

It was the *hua na fai*'s duty to determine how and where the irrigation systems would be constructed, and to organize villagers to contribute building materials and labour. He also calculated the amounts of water to which each farmer was entitled, and adjudicated water disputes. Each May (and sometimes December), before the planting season began, he would determine the repairs required and muster the corvée labour to carry them out. In addition, he collected fees to cover the costs of maintenance and of rain propitiation ceremonies.

Labour requirements for weir construction and repair varied from village to village but, in general, farmers owning 1–10 *rai* (0.4–4.0 acres) were required to supply one worker; two workers for 11–20 *rai*; three for 20–30 *rai* and so on. Workers had to provide their own tools and contribute their labour for ten days each year, usually just before the planting season when there was little work activity in the village.[22]

The *hua na fai*'s authority and the sacrosanctity of the weir systems emanated from taboos and spirit beliefs. Shrines were erected on river-banks to house the guardian spirits (*phi khun nam*) which protected the weirs. *Fai ton kaeo* and *phi khun nam* spirit propitiation ceremonies were conducted at the weirs where the spirit houses (*hu phi fai*) were located, and were often combined with rituals to encourage the gods to send down abundant rain. After the annual dredging of canals, normally in June, a black cow would be sacrificed to appease the weir spirits.

Underscoring the importance given to river management, the *hua na fai*'s responsibilities ultimately came to encompass a much broader sphere. As a village grew, the *hua na fai* acquired wider authority over its affairs and additional power to carry them out, becoming the king's representative and the virtual ruler of the village. Such extended authority enabled him to administer the rather harsh penalties meted out for violating *muang fai* ordinances.

The importance attached to irrigation water is evident in the system of regulations and fines governing its use. For example, one law stipulated that domesticated elephants could venture no closer than 50 *wa* (100 metres) from a canal. For minor infractions—refusing to join the corvée labour force, forgetting one's tools, accidentally breaking a weir, or allowing one's elephants or cattle to do so—villagers would be assessed moderate fines. The prohibition against destroying weirs extended to the spirits that guarded them. The Wat Saohai manuscript notes that 'whoever destroys the spirit house near the weir offends the spirits and can cause destruction of the weir'. Stealing water was regarded as a crime against society and the punishment suggests that this offence was not looked upon lightly: 'In such case if he [the thief] is killed by the neighbour, it is justified.' Were the thief to be caught by the community, he would be subject to a crippling fine assessed by the *hua na fai*.[23] Eventually, the punishment was reduced to being knocked on the head and, by the nineteenth century, to a simple fine.

The *muang fai* system persisted with minor modifications into the twentieth century. When the Royal Irrigation Department (RID) was established, it augmented and ultimately usurped the *muang fai* system's authority by building permanent barrages and irrigation channels which required minimal maintenance. The adoption of pump irrigation and piping technology also reduced the system's importance and the *hua na fai* became a simple village official under the authority of the village headman.

Although the *muang fai* system was pervasive among farmers in the northern hills and valleys, it found less favour among the hilltribes who chose to build temporary villages on the hills and pursue swidden (slash-and-burn) agriculture. Instead, tribesmen planted upland rice and other crops whose irrigation needs were supplied by the clouds. For bathing and domestic requirements, they hauled water uphill in jars or tapped a stream high up its course and dug a channel to convey it into village ponds.

Central Plains Irrigation and Agriculture

The embayment of a large portion of the present-day Central Plains and the marshy nature of the land bordering the bay would have rendered agriculture, and habitation, problematic in the lower basin. Even when the ocean receded and the marshes dried, successful cultivation would have been difficult given the salt deposits left by the falling sea waters, as is evident today in the salt flats near Samut Sakhon. The speed with which the alluvial fans moved southwards into the Bight of Bangkok suggests that the rivers carried heavy sediment loads from the north during the monsoon runoff, and that within a very short period, the bay was filled by fertile, albeit clayish soil. Even today, these sediments measure 15 metres deep and sit atop 1800 metres of silt and deltaic clastics.[24]

In overlaying its delta with rich topsoil carried from the hills, the Chao Phya resembles the Nile in its pre-Aswan Dam period, a kinship noted by numerous seventeenth-century European visitors to Ayutthaya, including Chaumont: 'The river is grand and graceful. It's several *wa* deep. The rice fields along the river are normally flooded 4–5 months a year.... The Chao Phraya River is similar to the Nile of Egypt in that flood waters carry natural fertilizer to the many fields along its banks.'[25]

Little is known of the earliest inhabitants of the Central Plains. There is evidence dating from 3000 BP of fields defined by bunds and of the use of buffalo as draught animals.[26] These indigenous people were joined between the fifth and tenth centuries by Mon–Khmers from southern Burma who settled along the riverbanks of the lower basin. Bringing with them a relatively well-developed civilization, they pursued a rice-based agriculture that relied primarily on rainfed irrigation.

As the quote from Chaumont suggests, the abundance of flood water and the techniques for using it may explain why Central Plains farmers did not expend the same effort in irrigation engineering that their cousins in the North did. The flatness of the lower plains mitigated against large-scale irrigation projects, and the lack of manpower and engineering skills available to harness the larger rivers precluded construction of elaborate irrigation schemes.

There may have been little need for them anyway. As the seas receded, the draining waters cut into the soft clays and created their own canals. A look at a map of the lower valley reveals thousands of kilometres of natural canals that web like tracery across the plains. Early inhabitants practised a relatively simple wetland rice cultivation that required little more than clearing forests, constructing bunds, and waiting for the rain to fall. As noted by Bennett Bronson, whereas in pre-agricultural times the Thai environment had been below average in capacity for supporting human populations, changing climate and geography had made it well above average in its ability to support crop cultivation.[27]

The largess of the Chao Phya River system supplied the foundation stones for the development of stable communities, providing both the fertile land the rivers laid down and the abundant water to irrigate crops. Already at this early date, rudiments of talent at manipulating streams of water to flow where needed were apparent in the *muang fai* system. In the next phase of Thai development—the successive kingdoms of Sukhothai and Ayutthaya—these skills would be employed on a grander scale to create the prosperity which would make Thailand an important economic power in South-East Asia.

1. W. J. Van Liere, 'Mon–Khmer Period from 200 to 1200 AD', in *Culture and Environment in Thailand: A Symposium Sponsored by the Siam Society*, Bangkok: Siam Society, 1989, p. 157.
2. John Crawfurd, *Journal of an Embassy to the Courts of Siam and Cochin China*, London, 1828; reprinted Kuala Lumpur and Singapore: Oxford University Press, 1967 and 1987, p. 187.
3. Phya Anuman Rajadhon, *Chiwit Chao Thai Samai Kon le Kan Suksa Ruang Prapheni Thai* [Thai Life in Olden Times and Studies of Thai Traditions], Bangkok: Samnak Phim Khlang Wittaya, 1972, p. 305.
4. 'Letter from HRH Prince Damrong Rajanubhab to HRH Prince Narisara Nuwattiwong of 5 November 1940', in *San Somdet* [Letters of Princes], Bangkok: Khurusapha Press, 1962, Vol. 20, pp. 40–1.
5. Nicolas Gervaise, *The Natural and Political History of the Kingdom of Siam*, Paris: Claude Barbin, 1688; English edn., trans. John Villiers, Bangkok: White Lotus, 1989, p. 16.
6. Crawfurd, *Journal of an Embassy to the Courts of Siam and Cochin China*, p. 438.
7. MR Seni Pramoj and MR Kukrit Pramoj, 'Letter to Phya Montri Suriwongse, the King's Ambassador to the Court of Queen Victoria, and Chao Mun Sarapethbhakdi, Vice-Ambassador, 1857', in *A King of Siam Speaks*, Bangkok: Siam Society, 1987, p. 212.
8. Anatole-Roger Peltier, *Pathamamulamuli: The Origin of the World in the Lan Na Tradition*, privately published, Chiang Mai, 1991, pp. 197–8.
9. The original inhabitants of the Lanna kingdom comprising the northern portion of Thailand.
10. Prinya Nutalaya and Jon L. Rau, 'Bangkok: The Sinking Metropolis', *Episodes*, Vol. 1981, No. 4, 1981, p. 4.
11. Douglas Anderson, 'Prehistoric Human Adaptations', in *Culture and Environment in Thailand: A Symposium Sponsored by the Siam Society*, Bangkok: Siam Society, 1989, pp. 110 and 114.
12. J. R. P. Somboon and N. Thiramongkol, 'Holocene Highstand Shoreline of the Chao Phraya Delta, Thailand', *Journal of Southeast Asian Earth Sciences*, Vol. 7, No. 1, 1992, p. 59.
13. J. R. P. Somboon, 'Coastal Geomorphic Response to Future Sea-level Rise and Its Implication for the Low-lying Areas of Bangkok Metropolis', *Tonan Ajia Kenkyu* [South-East Asian Studies], Tokyo, Vol. 28, No. 2, September 1990, p. 165.
14. Dhida Saraya, 'State Formation in the Lower Tha Chin-Mae Klong Basin: The Historical Development of the Ancient City of Nakhon Pathom', in *Culture and Environment in Thailand: A Symposium Sponsored by the Siam Society*, Bangkok: Siam Society, 1989, p. 181.
15. Thaweesak Jaritkhuan, 'A Study of Distribution in Sand Sediment Layers on the Bank of the Old Chao Phya River in Phra Nakhon Si Ayutthaya', M.Ed.

thesis, Srinakarinviroj University (Prasanmitr), Bangkok, 1990.

16. Wilhelm G. Solheim, 'The New Look of Southeast Asian Prehistory', *Journal of the Siam Society*, Vol. 60, Pt. 1, January 1972, pp. 1–20.

17. Bennett Bronson, 'The Extraction of Natural Resources', in *Culture and Environment in Thailand: A Symposium Sponsored by the Siam Society*, Bangkok: Siam Society, 1989, p. 292.

18. Sommai Premchit and Amphay Dore, *The Lan Na Twelve-month Traditions*, Chiang Mai: Toyota Foundation, 1992, p. 7.

19. Chit Phumisak, *Khwam Pen Ma Khong Kham Sayam, Thai, Lao le Khawm, Le Laksana Thang Sankhom Khong Chu Chon Chat* [The Origins of the Words Sayam, Thai, Lao, and Khom and Social Characteristics of These People], 2nd edn., Bangkok: Samnak Phim Duang Kamol, 1981.

20. Yoneo Ishii, 'History and Rice-growing', in Yoneo Ishii (ed.), *Thailand: A Rice-growing Society*, trans. Peter Hawkes and Stephanie Hawkes, Honolulu: University Press of Hawaii, 1978, p. 19.

21. Vanpen Surarerks, *Historical Development and Management of Irrigation Systems in Northern Thailand*, Chiang Mai: Department of Geography, Chiang Mai University, 1986, p. 82.

22. Ibid., pp. 128–9 fn. 3.

23. Ibid., p. 92.

24. Prinya and Rau, 'Bangkok: The Sinking Metropolis', p. 4.

25. Mr le Chevalier de Chaumont, *Relation de l' Ambassade à la cour du roi de Siam* [An Account of the Embassy to the Court of the King of Siam], reprinted Bangkok: Chalermnit Press, 1985, p. 42.

26. Phya Anuman Rajadhon, *Essays on Thai Folklore*, Bangkok: Social Science Association Press of Thailand, 1968, p. 134.

27. Bronson, 'The Extraction of Natural Resources', p. 295.

2 Liquid Road to Riches

When the Tai [progenitors of the Thai peoples] chose a site for a city or village the most important criterion was water, and the second was earth. Only after these two criteria satisfied them did they proceed to consider other factors, and finally settle down.[1]

LIKE most Asian leaders in the last millennium, Thai monarchs in North and Central Thailand built their principal towns beside rivers in the belief that flowing water was an integral component of urban design, providing sustenance, mobility, and protection. In so doing, the Thais melded two worlds, the liquid and the solid, into a cohesive whole, making island fortresses of their cities and roadsteads of their waterways.

Their initial relationship with rivers, however, was extremely tentative. The choice of site for Chiang Mai, the capital of the Lanna kingdom, the pre-eminent power in the north in the thirteenth century, exemplifies a fear of the river's power, a timidity in confronting it, a hesitation founded on hard experience. Sometime after the fourth century, Chiang Mai's first inhabitants, the Luwa (Thai Lue), built a small settlement on the far western edge of the valley at the foot of Doi Suthep. Some time later, they moved nearer the river to Chiang Mai's present site, erecting the Inthakhin pillar which now stands at Wat Inthakhin, also called Wat Sadu Muang (Navel of the City). For unknown reasons, King Mengrai, considered Lanna's greatest monarch, rejected this site, and in 1286 chose another to the south-east on the right bank of the Ping. Near the present Wat Chedi Si Liem, he constructed the Wiang Kum Kham palace on the ruins of another old Luwa settlement.

At that time, the Ping flowed just east of the palace along the Rong Ping Hang (Old Ping Channel) that now parallels the Chiang Mai–Lamphun Road. In the following decade, the river changed course. The Lanna Chronicles record that it flooded the palace compound and buried nearby Wat Pa Tan in silt. In 1296, recognizing the impracticality of a riverside location, King Mengrai took up residence on the high ground of Wat Chiang Mun, 1 kilometre west of the river, and began planning a new royal city. According to tradition, he invited King Ramkamhaeng, ruler of the neighbouring kingdom of Sukhothai, to advise him on its design. What Ramkamhaeng offered his contemporary was a seven-point geomancer's formula (*nimit chaiya mongkhon chet prakan*) whose

THE CHAO PHYA

derivation is unknown, but which may have played a role in Sukhothai's design.

Two of the formula's four main criteria relate to rivers. There must be a mountain west of the city, sloping down to the east (in Sukhothai: Khao Luang; in Chiang Mai: Doi Suthep), a dammable stream issuing from a nearby hill (Sukhothai: Saritphong Dam; Chiang Mai: Wat Fai Hin [Temple of the Stone Dam]); a river east of the town (Sukhothai: Yom River; Chiang Mai: Ping River), and a divinity area 3 kilometres from the western city wall at the foot of the mountain (Sukhothai: Aranyik Park; Chiang Mai: Wat Suan Dok, Wat Umong, and Wat Pa Daeng).[2] Following these dicta, Mengrai built a fortified city surrounded by a wall and a moat. Unlike England's King Canute, Mengrai recognized the folly of challenging flowing water and wisely chose not to transfer the old city's name, Kum Kham (Control the Water, in northern Thai), but settled instead for the neutral Nopburi Sri Nakhon Ping ka Chiang Mai (New City on the Banks of the Ping River).

Because Chiang Mai stood at a higher elevation than the Ping, its moat was filled, not by the river, but by water channelled from the Wat Fai Hin (Plate 6). The task of filling the moat was complicated by the slope of the land; the north-western corner where the canal enters the moat stands 316 metres above sea-level, and the north-eastern corner, 1.6 kilometres to the east, is 9 metres lower.

6. The moat surrounding Chiang Mai is filled by a stream issuing from Wat Fai Hin on the flanks of Doi Suthep.

LIQUID ROAD TO RICHES

Lanna engineers overcame the problem of the gradient by installing a series of locks along the city's northern and southern walls to create ponds. The locks are still in place but are now overlaid by streets which carry traffic in and out of the city (Plate 7).

Even at this early date the river seems to have played a vital role in northern daily life. A study of the 'Customary Law of Kings Ku Na and Mengrai' reveals that the northern rivers were swarming with boats. The earliest entry, dating from 1238, concerns a judgment made when a heavily laden boat collided with a light boat. Another records a damage claim brought against the owner of a vacant, drifting boat.[3] The Ping River was also the principal conduit for communication with the southern cities of the Lanna kingdom. Located at the Ping River rapids called Kaeng Soi, halfway to the southernmost Lanna town of Raheng (Tak), was the sentinel city of Wiang Soi Si Suk (Plate 8). The Lanna Chronicles record that its control of trade along the river earned it sufficient wealth to finance the construction of 99 *wat*. For unknown reasons, it declined in importance and was ultimately abandoned. After 1964, the rising waters of the Bhumibol Dam reservoir covered the ruined city. Some artefacts were recovered, but today only a restored *chedi* on the western bank stands as testament to its existence.

7. To compensate for the steep gradient along its length, the Chiang Mai moat was fitted with a series of locks to regulate water-levels.

8. All that remains of the northern sentinel city of Wiang Soi Si Suk (*right*) is this restored *chedi* (*above*). It rises above the former Kaeng Soi rapids which once hindered passage along the Ping River south of Chiang Mai but which now lie buried beneath the waters of the Bhumibol Dam reservoir.

Sukhothai's Dual Cultures

Chiang Mai's contemporary, Sukhothai, lay in the Yom River Valley about 300 kilometres to the south-east. The valley appears to have been inhabited by two cultures, Thai and Khmer, who lived side by side and practised two distinct types of farming. Sometime before the twelfth century, Thai–Lao immigrants had been filtering south from the northern hills to settle the upper portion of the Chao Phya Valley, notably along the Yom River just north of Sukhothai. Although this portion of the valley floods to considerable depth, the Thais turned a liability into an asset, utilizing its waters to irrigate their rice crops. As Van Liere notes, the Thai–Lao

> brought with them a novel idea in water management. They breached the riverbanks of major rivers at many places, so that the backswamps functioned as regulators against the rapid rise of river levels. Towards the end of the rainy season, they drained the backswamps quickly, so that receding flood agriculture could be practiced on a larger scale. These breaches can still be seen at many places ... along the Yom, down from Si Samrong, where the alluvial river complex of riverbanks and backswamps, becomes very wide.[4]

Like many Asian peoples, the Khmers, who sometime before the thirteenth century founded Sukhothai as the north-western outpost of the Angkorian empire, seem to have distrusted rivers. In

the dozens of cities in their vast realm, they put as much distance between themselves and the rivers as possible, staying only as close to them as necessary for fishing and transport purposes. Instead, they built their towns on high ground well above the rivers, and channelled water from nearby mountains for household and religious purposes. On the natural terraces between a town and a river, they transplanted glutinous rice in bunded fields which were irrigated both by rainfall and by standing water which spilt from field to field. In establishing Sukhothai, they followed the same principles, siting the city some 12 kilometres west of the Yom River.

On first consideration, the choice of a site for Sukhothai seems strange given the paucity of rainfall to irrigate crops. Contrary to the general contention that the Sukhothai Valley is a lush rice basket, it lies in the rain shadow of the western mountains and suffers one of the lowest rainfalls in Thailand, an average of 643 millimetres per year. This factor would seem to mitigate against success in farming and development but studies of archaeological sites throughout South-East Asia reveal that most large towns were situated in areas of low, seasonal rainfall.[5] Thus, the choice can only be explained by the weight of tradition and by the Khmers' apparent preference for a rice variety with low water requirements. As a counterbalance, Khmer towns escaped the annual floods that have always plagued Chao Phya Valley cities.

Little is known of Sukhothai's early irrigation works. Theories often conflict as to whether a particular construction was the work of the Khmers or their Thai successors. It appears that the city was served by a small stream that rose in the mountains to the south-west, flowed across the terrace, and then drained down the valley slope into the Yom. It is also well known that Sukhothai's Khmers excavated numerous large stream-fed *baray* (ponds) which served religious purposes. Household water presumably came from the stream itself or from wells.

Early in the thirteenth century, Khmer hegemony in South-East Asia began to wane, weakened by internal conflicts at Angkor. After constructing three religious buildings, the Khmers withdrew to their empire's original borders, leaving Sukhothai to the Thais, who selected it as the first capital of their new nation. Most modern knowledge of the Thai portion of Sukhothai's history is derived from the so-called Ramkamhaeng Stone inscriptions whose authenticity has recently been called into question. They record that in 1287, King Ramkamhaeng began construction of the Saritphong Dam across the south-western stream, and completed it in 1327. Restored in the 1970s by the Department of Fine Arts, the dam created a reservoir with a capacity of 380 000 cubic metres. All evidence suggests that the Thais used the reservoir as a settling pond to remove sediment from the water and thereby render it potable. Given its small size, the dam probably did not serve irrigation purposes.[6]

The inscriptions also relate that the Thais built a *muang fai* scheme on other streams, perhaps to supply water to the city and to irrigate the field crops planted on the terraces. Inscription III, dated to 1357 in the twilight years of Sukhothai's power, notes that 'Phraya Thammaraja [King Maha Thammaracha I (r. 1347–70)] ... was versed in the skills of *muang* digging and *fai* construction'.[7] A *muang fai* system alone would not have supplied sufficient quantities of irrigation water for the large-scale cultivation required to support Sukhothai's population and its political aspirations. If the Yom was not employed for the purpose—and assuming that the rice crop was not entirely rainfed—the water must have been conveyed from another source. Until recently, there were few clues as to what that source might have been.

For decades, scholars have been aware of an ambitious construction project undertaken early in Sukhothai's history. Earthworks were raised for what appeared to be a road (Plate 9) that began at the city of Si Satchanalai, an important Thai satellite town on the banks of the Yom River, 55 kilometres north of Sukhothai. It ran south through Sukhothai, and a further 68 kilometres south-west to its terminus at Kamphaeng Phet, a fortified city on the Ping River that also defended the kingdom. On completing a survey of the earthworks in 1907, King Chulalongkorn (Rama V) named it Thanon Phra Ruang (Phra Ruang Road), Phra Ruang being a title applied indiscriminately to all eight of Sukhothai's early monarchs.

The so-called 'Inscription XIII' on the base of a Shiva image dating from 1510 and found at Kamphaeng Phet describes the renovation of the city in the early sixteenth century and offers an intriguing counter-explanation for the 'road'. The passage notes:

Since the *thaw* [channel or pipe] which had been built purposely by Phu Phraya Ruang ... had become silted up, our people had to depend solely

9. The substrata of the Phra Ruang Road which some historians have suggested was a canal serving Sukhothai.

LIQUID ROAD TO RICHES

upon rain water for the rice fields. But now that the *thaw* has been discovered and dredged, they no longer practice sky farming [depending solely on rainfall] but irrigated rice cultivation.[8]

The inscription suggests that it was a canal, not a road, that Phra Ruang had built.

An examination of the region's terrain supports such a contention. Topographical maps reveal that Kamphaeng Phet lies 79 metres above sea-level and Si Satchanalai, 66 metres; Sukhothai stands only 49 metres above sea-level. The greater height of the two outer cities would have provided sufficient gradient for water to flow by gravity along a canal from the Ping at Kamphaeng Phet, and another at the Yom at Si Satchanalai, to the capital (Map 4).

If, indeed, it was a canal, time and erosion have erased much of it. Aerial photographs of its remains reveal that in several portions two embankments run parallel a few metres apart. This is an odd arrangement for a road, which requires only a single track, but could indicate dikes bracing a canal. Moreover, the twin lines reach the walls of Sukhothai, not at the central portion where the city gates were located—as would be expected were it a road—but at the south-west corner of the city wall, as if entering the moat which, it is well known, surrounded the city.[9]

A canal, or combination canal–road, would have aided Sukhothai kings in maintaining control over their realm (Plate 10). Although Chaliang on the Yom River, north-east of Sukhothai, had been the principal Khmer administrative centre in the twelfth century, it was eclipsed by Sukhothai in the thirteenth century. Sukhothai's Sri Intharathit dynasty, struggling with the Khmers to establish itself as the dominant power in the region, would have been aided immeasurably by the canal, which together with its branches would have permitted the cultivation of new lands, thereby adding to the kingdom's wealth.[10]

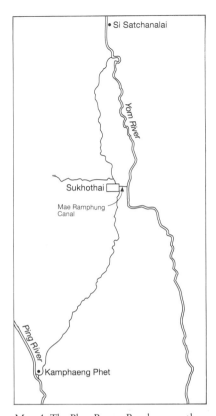

Map 4. The Phra Ruang Road ran north from Kamphaeng Phet to Sukhothai, and then to Si Satchanalai.

10. The moat of Kamphaeng Phet formed a partial defence against enemies.

27

The canal theory has been challenged on a number of grounds, the most notable being that while the land slopes towards Sukhothai from Si Satchanalai and Kamphaeng Phet, it undulates with elevations fluctuating between 25 and 30 metres. Such topography would not affect a road but because a canal must maintain a steady gradient along its entire length, engineers would have been forced to curve it around hill contours, or to construct massive earthworks to raise it above the low spots and maintain the gradient. The canal theory's supporters reply that the engineers did indeed build levees to raise the canal but that these supporting walls crumbled over time. Corroborating evidence is needed to buttress both arguments but it is interesting to note that, in 1987, when His Majesty King Bhumibol Adulyadej (Rama IX) surveyed a route for an irrigation canal running north from Kamphaeng Phet, it is the route of the Phra Ruang Road that he chose.

While Sukhothai's hydraulic engineers gained considerable experience in dealing with subtle variations in gradient along the small tributaries, they were not prepared to brave the river itself. It remained for Ayutthaya, the next major political power in the basin, to rise to the challenge, doing so on such a grand scale that its rulers elevated Siam (as Thailand would be known until the twentieth century) to the status of a major power in South-East Asia.

It is interesting to note that the planners of modern Sukhothai ignored the lessons of history and erected their new city on the banks of the Yom. Each year when the rain-swollen river rises, it floods the town, damaging property and wrecking havoc on city life.

After the fourteenth century, an improved knowledge of hydraulic technology emboldened Siamese monarchs to challenge the rivers. Henceforth, they would site their cities on river-banks and pursue agriculture in the flood plains, planting their rice with the first rains and waiting for the rivers to overflow their banks and inundate the rice fields with life-giving water. In succeeding centuries, rivers would serve as highways to link towns; canals would serve as town streets; and rivers would be channelled to form defensive rings around cities, in part to compensate for the lack of good stone with which to build walls. Most Thai city walls were built of brick or laterite, vulnerable to an enemy bent on burrowing beneath them. Brick walls rising steeply from a moat denied an enemy the foothold it needed to begin the work of tunnelling.

Several Thai cities were split by rivers running through their hearts, an open invitation, it would seem, to water-borne invasion. These towns are often called *ok taek* (broken chest) to describe a fortification wall broken where the river runs through it, dividing the city into two halves like Hungary's Buda and Pest. Phitsanulok and Suphan Buri are often referred to as *ok taek* towns. Other historians argue that Thai towns were originally walled fortresses. As they grew, new walls were built to surround a new section, eventually creating hive-like cells, each unique to itself and independent from its neighbours. Aerial views of sites dated to 1500 BC reveal an utter lack of uniformity of design, with dozens of shapes ranging from geometric to amoebic. Thus, the so-called *ok taek* may be misnomers for early cities whose riverside walls were later pulled down when enemies no longer threatened.

Ayutthaya's Rise

At the height of its power, Sukhothai's southern boundary ended at Chainat. Various theories have been postulated for the kingdom's decline: an epidemic, insufficient supplies of water to support a growing population, a preoccupation with religion to the neglect of statecraft. One key limiting factor was its lack of river access to the sea and, therefore, its inability to control the realm's southern portions or to benefit from trade. Its upriver rival, Si Satchanalai, conducted a thriving trade in ceramics produced at its Sawankhalok kilns and shipped to China via the river and the sea. Sukhothai was essentially land-locked. It was left to a new kingdom to establish a presence lower down the Chao Phya that would enable it to control river traffic and levy taxes up and down the river (Map 5). Within a few short years, that city, Ayutthaya, would become the premier political power in Thailand.

Like Brasilia and other planned cities, Ayutthaya was carved from the jungle. Unlike them, it owed its genesis to river engineers. Prior to 1350, Suphan Buri and Lopburi had been the principal population centres of the lower basin. Driven south by a smallpox epidemic at Suphan Buri, Prince U-Thong and his army marched several days east until, according to the Royal Chronicles, they reached the banks of the Lopburi River. Crossing it, they camped on Dong Sano, a 'circular island, smooth, level, and apparently clean'.[11] Dong Sano was protected by a double ring of water; it was an island in the middle of a large pond on a lobe of land embraced by an oxbow of the Lopburi River. Here, U-Thong's armies took up residence, declaring the island the capital of the new kingdom of Ayutthaya, so named for the holy city of Ayodhya in India.

Thus, from the beginning, Ayutthaya's destiny was written in water. Its Sanskrit name, Krung Thep Dvaravati Si Ayutthaya, is derived from the Mon word *krerng*, meaning 'canal' or 'river'.[12] Although located in the heart of a fertile rice region, it sat on level ground, open to enemy attack. U-Thong compensated for this disadvantage by turning his city into a mini-Mont-Saint-Michel, a citadel ringed by a wall of earth and water, much like the mythical Phrasumen Mountain which, according to Hindu cosmology, occupied the centre of the universe. Ayutthaya's distance from the river-mouth protected it from sea-borne invasion. The rivers and canals which radiated from it were arteries for transport and were its principal avenues of contact with a further world.

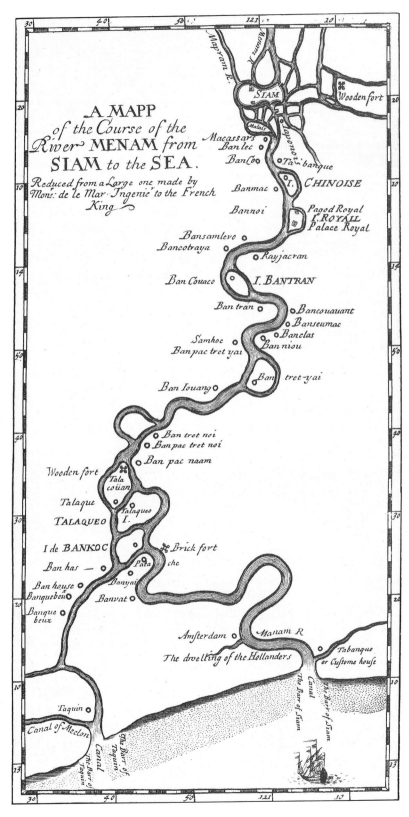

Map 5. Ayutthaya and the course of the Chao Phya to the Gulf of Thailand, 1680s. (la Loubère, 1693, facing p. 5)

As noted, Ayutthaya and its island were embraced by a loop of the Lopburi River extending to the west. The Pasak River flowed along the eastern perimeter of the city, a few kilometres east of its present course, and joined the main river south of Ayutthaya. When he declared the city as his capital in 1351, U-Thong ordered his army to dig a short shunt which carried the waters of the Pasak into the Lopburi above the city. He then dug a 3.5-kilometre north–south channel to form the city's eastern boundary, effectively creating a moated fortress measuring 4.0 by 2.5 kilometres.

During its first two centuries, Ayutthaya was involved in numerous territorial wars with Sukhothai and northern neighbours, and the river played a role in military strategy (Plate 11). When Sukhothai's King Maha Thammaracha III (r. 1398–1419) seized Nakhon Sawan from Ayutthaya in 1400, he succeeded in 'blocking riverine communications at a critical junction', allowing him time to conquer Nan and Phrae.[13] In time, however, Ayutthaya grew powerful, and by the mid-sixteenth century, held sway over territory bordered on the north by Si Satchanalai and on the west by the Bay of Bengal.

The kingdom's economy was founded on wetland, transplanted rice nourished by the river, a species that originally grew wild in the lower reaches of India's Ganges, Bangladesh's Brahmaputra, Burma's Irrawaddy and Salween, and Indo-China's Mekong River basins. The fertility of the Chao Phya ensured yields high enough to supply domestic consumers, with considerable surplus available for export. In this respect, the river served not only as a rich medium for rice cultivation but as a highway to overseas markets.

11. The Chao Phya as a moat which provided a protective wall around Ayutthaya until it was breached by Burmese armies in 1569 and 1767.

Prized by foreign consumers, Ayutthaya's savoury rice was shipped downriver for sale in Malacca, on the west coast of Malaya, whose merchants presumably resold it to buyers elsewhere. After several decades, buyers began bypassing the middlemen, sailing to Ayutthaya to make their own purchases. Wyatt notes the impact of this trade in strengthening Ayutthaya's position in the region:

> The effects of economic development, spurred by dramatic increases in international trade in the fifteenth and sixteenth centuries, on the rise of Ayutthaya cannot be understated. They worked at times in an almost circular fashion. The more the king gained wealth through trade, the better able he was to overawe or overcome both domestic and neighboring rivals and join their territory to his, thereby further improving his ability to trade.[14]

The first foreign merchants in Ayutthaya were Asian, but by the sixteenth century, European galleons were sailing up the Chao Phya (Colour Plate 3). The Portuguese were the first to drop anchor outside Ayutthaya's walls, arriving in 1511, and followed soon after by Dutch, British, and French ships.

With the new wealth earned from foreign trade, Ayutthayan rulers began building an empire. As the aggressors, they were little concerned about erecting defensive walls around their island city, trusting the river to protect them. By the mid-sixteenth century, however, a new enemy—the Burmese—appeared on the northwestern borders and Ayutthaya built earthen ramparts to bar their entry. After conquering Chiang Mai, the Burmese made a foray down the Ping River as far as Tak and then, in 1548, struck for Ayutthaya. Siam's King Maha Chakraphat (r. 1549–69) sallied forth to confront them at Lopburi, Nakhon Nayok, Phra Padaeng, and Suphan Buri. The Burmese broke through at Suphan Buri but were slowed by streams and rivers. Shortly after, their invasion was blunted by Siamese marines attacking from boats.

Reinforced by Lanna soldiers, the Burmese renewed their attacks, advancing to the walls of the city where the Siamese put them to rout. The Hmannan Yazawin Dawgyi (a history of Burma, compiled by order of King Bagyidaw of Burma in 1829) records that

> The invading army then followed up but was unable to take the city owing to its commanding position, surrounded as it was by water, and also to the strong defense made, in which the artillery served by Kala Panthays [foreigners] played an important part in keeping the invaders off at a safe distance.[15]

The Burmese called off their siege and returned home. Shaken by the attacks, King Chakraphat dismantled Ayutthaya's original earthen walls and replaced them with stout brick walls. Having seen the strategic value of a navy, he built a fleet of warships.

Between 1549 and 1592, the Burmese would return six times to besiege the city. In the invasion of 1564, the river played a key role in defeating them. British merchant Samuel Purchas wrote that the Burmese,

> seeking politike delayes, made semblance still to deliver, until in the third moneth after the River overflowed the country six score miles about, after his yerely custome, and partly drowned, partly committed to the Siamites (attending in boats for this booty) to be slaughtered, that huge army; of which, scarce three score and ten thousand returned to Martovan [Martaban].[16]

In 1568, the Burmese reappeared. At first thwarted by the river-moat, they dug trenches up to the river to protect their soldiers and then attempted to build bridges, probably on the backs of boats, to cross to the city walls. They failed and settled into a siege interrupted by artillery duels. After numerous incursions, the Burmese finally took the city in 1569 after treacherous Thais reputedly opened the gates. The Burmese soon withdrew to repel enemy invasions into their own territory, and Ayutthaya was left to pick up the pieces. It recovered quickly despite six invasions by the Cambodians between 1570 and 1587, and by 1580 had rebuilt its city walls. In 1590, King Naresuan constructed a fleet of *rua chai* (victory ships) whose bows were fitted with cannons. *Khrut* (garudas) and other fierce beasts were carved into the bows to instil fear in the enemy (Colour Plate 4). Employed in battle against King Honsaowadee of Burma, the 'battleships' were instrumental in halting the invaders at the banks of the Tha Chin River.

Looking Westward

That Ayutthaya became a major Asian entrepôt is testament to its skills as a mercantile power since it lay 140 river kilometres from the sea. Except at high tide, a river-mouth mudbar prohibited entry to ships larger than 400 tons or with draughts of more than 3.5 metres.[17] Although the twisting, narrow river prevented large sailing ships from making long tacks, the value of Thai rice, deer and other hides, minerals, and other products offered sufficient inducement for them to make the arduous journey upriver. Gervaise noted that 'the mines of tin, iron and saltpeter, the cotton, silk and perfumes that are found in abundance in this kingdom could make it the richest state in the Indies'.[18] La Loubère added that 'Campeng [Kamphaeng Phet] is known by the Mines of excellent Steel'.[19] Europeans also came for gold, silver, and precious stones. Other Siamese products prized by both Europeans and Asians were rice, spices, and forest products including sappanwood (Brazil Wood, used for dyes), eaglewood (an aromatic wood used as incense), lac (a resin for shellac), and benzoin (a tree resin for perfume).

In turn, Ayutthaya imported Indian cloth, luxury goods, firearms, and metals. These early cargoes flowed along the Chao Phya in the holds of Chinese and European ships. After a time, Siamese craftsmen utilized skills learnt from Chinese carpenters to build and man the ships that carried Thai commodities to overseas markets. By the reign of King Narai, Ayutthaya was regarded as one of the most prosperous nations in Asia.

Seventeenth-century Ayutthayan geography and economy is portrayed in graphic detail in European merchants' journals, most of which were written as commercial intelligence for their trading companies in Europe. As the only reliable Siamese manuscripts that survived the eighteenth-century destruction of Ayutthaya—the Loeng Prasert Chronicles and the Royal Chronicles of Ayutthaya—deal primarily with royal successions and battles, these foreign accounts provide a detailed picture of daily life in Ayutthaya. The two foreign observers credited with presenting the most accurate portraits of the kingdom—the Frenchmen Simon de la Loubère and Nicolas Gervaise—wrote their accounts in the same year, 1688, at the zenith of the foreign presence in Ayutthaya.

La Loubère described the city thus:

It has almost the figure of a Purse, the mouth of which is to the East, and the bottom to the West.... The King's Palace stands to the North on the Canal which embraces the City; and by turning to the East, there is a Cause[wa]y, by which alone, as by an Isthmus, People may go out of the City without crossing the water.... Most of the streets are watered with straight Canals [Map 6], which have made Siam to be compared to Venice and on which are a great many small Bridges or Hurdles, and some of Brick very high and ugly.[20]

Sumet Jumsai depicts the watery network in detail in his book *Naga*, and in the process, conveys a sense of the canals' importance, the massive hydraulic engineering required to build them, and the sophisticated social organization which permitted kings to excavate and maintain them:

The city wall, excluding the citadel walls of the palaces, was approximately one kilometre long and equipped with seventeen forts. The Royal Palace had, in addition, seven forts and the Palace to the Front. There were nearly 100 city gates with at least twenty-two more if those of the Royal Palace were included, and of these about twenty were water gates to let the boat traffic through. The canals within the confines of the wall were 56.4 kilometres long and some twenty-eight permanent bridges spanned them.... The complete canal network, including suburbs, measured over 140 kilometres.[21]

Ayutthaya's palaces and *wat* were constructed on artificial knolls created from soil excavated from the lakes and canals, and most residences were built on stilts to raise them above the floods. The houseboats, which would later fill Bangkok's rivers and canals and elicit the wonder of foreign visitors, seem to have been confined to the side canals on the outskirts of Ayutthaya. They may have been prohibited on the narrow river because their presence would have obstructed navigation, sheltered invaders, and

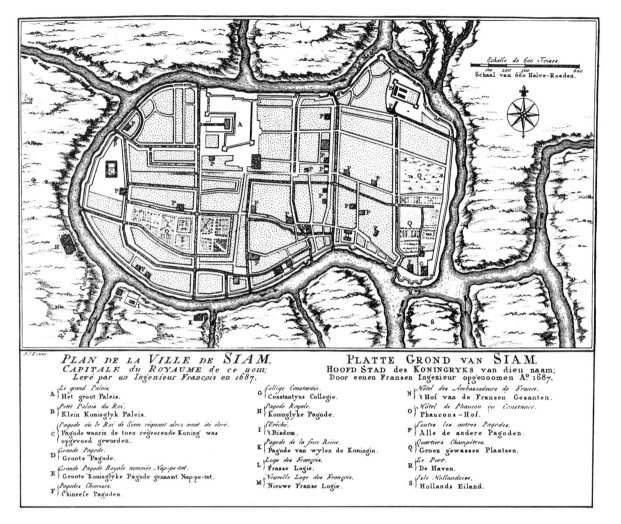

Map 6. A French map of Ayutthaya, 1687, details the various canals which formed the city's streets. (Prevost, 1765, Vol. XIV)

denied defenders on the walls a clear view of activities on the opposite banks.

The sea-like impermanence of the city is evident in many Ayutthaya-period *wat* whose bases were incised with deck lines in the belief that the buildings were ships carrying the faithful to salvation (Plate 12). Given the low elevation of the surrounding land, it is probable that they were more than symbolic ships; they may virtually have been afloat when the annual floods inundated the city.

The harbour was located on the main river just south of the town, in the sector where foreigners were permitted to reside. From these enclaves, the Dutch, English, Portuguese, and French merchants jostled for favour in the Ayutthayan courts. They built warehouses, described as 'factories', to store goods for export while awaiting the arrival of company ships on the pre-monsoon winds. By situating the foreign harbour downstream from the palace, the

12. A deck line on an Ayutthaya-period *wat* at Wat Bang Kluai, north of Bangkok.

Siamese attempted to keep foreign pollution far from the palace located in the city's north-western quarter. 'Out of respect for the king's palace as a principal focus of the city, foreign nationals were in the past granted land downstream for their settlements.'[22] The pattern would later be repeated at Thonburi, with ships moored near the Portuguese Santa Cruz Church (Colour Plate 5), and at Bangkok with ships anchored in the area of the present Oriental Hotel. The selection of Khlong Toey as the main port during the twentieth century would be for practical reasons: the deeper water permitted berthage of deep-draught ships and eliminated the need to build upstream bridges with movable central spans.

Contemporary foreign writers exulted at Ayutthaya's splendour, and were particularly impressed by the wealth reflected in its golden pagodas which, depending on the writer, numbered up to 2,000. Beyond the city walls, however, stretched a vast sea of tangled green jungle. The only viable means of traversing it were the dozens of rivers which linked a half-dozen minor kingdoms. Gervaise noted that 'in these regions there are only inaccessible forests and vast wilderness. If by chance there happens to be a small hamlet, it will be inhabited only by some poor wretches who have taken refuge there so as to escape punishment for the crimes they have committed in their own country.'[23]

The Ayutthayan kingdom was bounded on the north by the riverside towns of Me-Tac (Tak), Lacontaje (Lampang), and Porselouc (Phitsanulok), with Sukhothai looking across to the Yom River. Phitsanulok was described as the last town before the frontier with 'Laos', which is to say, the Lanna kingdom with Chiang Mai, situated fifteen days' journey north of the border. Other important river towns included Suphan Buri, Louveau (Lopburi), 'Tian Ting [an old town, between Tak and Kamphaeng Phet, which disappeared], Campeng-pet [Kamphaeng Phet] ... Laconcevan

[Nakhon Sawan], Tchainat [Chainat], Talacoan [Nonthaburi], Talaqueou [Talat Noi, formerly located north of Samsen but no longer in existence], and Bancok'.[24]

While there are comprehensive lists of the kingdom's resources, there is general confusion over its geography. According to la Loubère, the perplexity was shared by foreigners and Thais alike: 'Navigation has sufficiently made known the Sea Coasts of the Kingdom of Siam, and many Authors have described them; but they know almost nothing of the Inland Country, because the Siamese have not made a Map of their Country, or at least know how to keep it secret.'[25] Contrary to the notion that topography was a state secret, it has been the experience of the author while paddling down the rivers that most Thais display little knowledge of, or interest in, geography or distances, even in their immediate vicinity.

In most foreign accounts, geographical knowledge of the northern region is glaringly faulty, appearing to have been based on conjecture rather than observation. Most reliable is la Loubère because 'that which I here present is the work of an European, who went up the Menam, the principal River of the Country, to the Frontiers of the Kingdom; but was not skilful enough to give all the Positions with an entire exactness'.[26] He correctly describes the upper rivers in this manner: 'At the City of Laconcevan [Nakhon Sawan] the Menam receives another considerable River which comes also from the North, and is likewise called Menam....'[27] He is, however, unaware of the role of the Wang and the Yom in the system and the region above Sukhothai is *terra incognita* among European and Siamese writers. Gervaise fails to identify his source as European or Siamese when he states:

Some people believe that this river is a branch of the Indus, others that it springs from the mountains bordering China and Laos. It seems more likely that it comes from a great lake which was discovered some years ago in Laos [meaning Lanna]. The Siamese generally accept this idea and even believe that the waters which flood their country each year originate from there.[28]

La Loubère is more cautious.

The Siamese do say that the City of Chiamai [Chiang Mai] is fifteen days journey more to the North, than the Frontiers of their Kingdom, that is to say at most, between sixty and seventy Leagues, for they are Journeys by water, and against the Stream.... Our [French] Geographers make it to spring from the Lake of Chiamai; but it is certain that it hath its source in the Mountains, which lye not so much to the North as this City.... The Siamese which were at that expedition [a military venture conducted 30 years earlier], do not know that famous Lake ... which makes me to think either that it is more distant than our Geographers have conceived, or that there is no such Lake.... However it may be doubted, whether the Menam springs from a Lake, by reason it is so small at its entrance into the Kingdom of Siam, that for about fifty Leagues, it carries only little Boats capable of holding no more than four or five Persons at most.[29]

That the Siamese documents fail to corroborate or deny his statement suggests that they may also have been uncertain about where the river rose. Gervaise relates a tale, which he does not attribute to a source, that 'some men who were sent out formerly by order of the king to find the source of the river, which still remains undiscovered, having traveled a long way in their search, were greatly astonished to find themselves back again at almost the same spot as that from where they had started.'[30] It should be noted that no northern lake of any size could exist given Lanna's rugged topography.

It is tempting to give credence to la Loubère's suspicion that the Siamese preferred to keep their knowledge of Thai geography 'secret'. Perhaps the Siamese feared the information could fall into enemy hands, so consigned it to memory to keep it from outsiders. It should be noted that Thais have always placed more reliance on oral than written communication. From the number of military expeditions launched into the North, it would appear that commanders had a thorough knowledge of the region. Today, even the simplest boatman can manœuvre his way unerringly through the massive web of canals and rivulets that looks to the outsider like an undifferentiated jungle maze.

Thus far, descriptions of the river have suggested it served military and commercial purposes but was otherwise devoid of traffic. In truth, the river teemed with boats. La Loubère introduces a description of the grand barges maintained by the king by saying that 'in this Country they go more by Water than by Land'.[31] Gervaise follows it with a description of water-borne commerce by ordinary Siamese: 'The majority of the population is engaged in trade. Some spend all their time trafficking on the river with their wives and children in large boats ... from which they almost never disembark.'[32]

River Engineering

To facilitate transport in the central valley, the Siamese initiated numerous canal excavation projects. A close look at a map of the region in the immediate vicinity of Ayutthaya reveals myriad canals whose die-straight lines indicate a human, not Nature's, hand in their creation. Their sheer numbers, running into the thousands, suggest that when the Siamese were not gripping a boat paddle or a plough, they were wielding a *jop* (hodag) or other digging implement, gouging out a shunt here or dredging a canal there. By the sixteenth century, Siamese engineers were modifying the Chao Phya itself. On the main river, canals were dug to serve three purposes. *Khlong rop muang* (canal around the city) served as city moats to bar enemy invaders. *Khlong chuam maenam* were designed to link one river with another. *Khlong lat* were channels dug across the oxbows of meandering rivers to

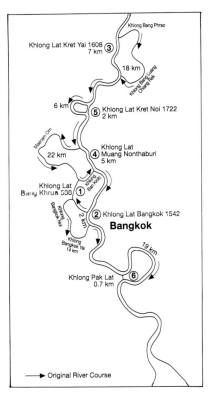

Map 7. The six *khlong lat* of the Chao Phya.

reduce travel distances. It was the creation of *khlong lat* at which Ayutthayan engineers excelled.

As international trade increased, Ayutthayan monarchs recognized the need to shorten transport distances between the capital and the sea. Although they were unable to dredge the bar to admit deeper draught ships, they realized they could speed the journey north by digging *khlong lat* in the winding portions of the lower river. Between 1538 and 1722, they dug six short *khlong lat* that lopped 62.3 kilometres from the length of the Chao Phya between Ayutthaya and the Gulf of Thailand (Map 7). The first *khlong lat*, dug in 1538, was a 3-kilometre shunt called Khlong Lat Bang Khrua that connected Wat Chalom (on what would henceforth be called Maenam Om) with Wat Khee Lek (on what would become Khlong Bangkok Yai) on the northern edge of present-day Bangkok. It shortened the journey from 9 to 6 kilometres but straightened the channel for easier passage. After 1636, it would be superseded by Khlong Lat Muang Nonthaburi.

The second canal would have a profound effect on the course of Thai history. There is disagreement about the exact date of its construction, with Gerini stating it was dug in 1538 and Nid Hinshiranan, a modern city planner, contending that it was dug in 1542. Its importance, however, is undisputed because it was responsible for the creation of Bangkok and Thonburi. The 2-kilometre-long Khlong Lat Bangkok (Plate 13) was dug from the site of the present-day Bangkok Noi Railway Station to a point just south of Wat Arun. River action widened it to become the main channel of the river, transforming the former loop into four canals—Bangkok Noi, Bang Ramat, Taling Chan, and Bangkok Yai—and reducing a journey of 14 kilometres to 2.

The third, Khlong Lat Kret Yai, was dug above Pathumthani in 1608 to shrink 18 kilometres to 7. In 1636, a dozen kilometres downriver, Khlong Lat Muang Nonthaburi was cut across the neck of Maenam Om by King Prasat Thong. It pared 17 kilometres from the 22-kilometre journey. Khlong Lat Kret Noi at the town of Pak Kret was dug in 1722, reducing the journey from 6 kilometres to 2.[33] The sixth *khlong lat* was below Bangkok at Ratburana, just south of Khlong Toey. A 600-metre-long canal called Khlong Pak Lat was cut across a narrow neck, effectively cutting 19 kilometres from the journey. Unfortunately, because the Chao Phya is tide-affected, the canal introduced saline water into the upper river, damaging marine and river-bank life. A dam would be built across its mouth in 1784 to halt salt water intrusion. Today, of the six *khlong lat*, only this one has failed to become the main channel of the river.

Several ambitious *khlong chuam maenam* engineering projects were undertaken during the Ayutthaya period. The date for the excavation of Khlong Samrong is unknown but it was redug in 1498 during the reign of King Ramathibodi II. Broadening an existing natural canal to allow the passage of ships, it connected

THE CHAO PHYA

13. The most famous of the *khlong lat*, Khlong Lat Bangkok now flows past Wat Phra Kaew, the Temple of the Emerald Buddha, and other major city landmarks.

the Bangpakong River with the Chao Phya River below Khlong Toey. Nearer the capital, la Loubère notes that 'the Siamese have cut a great many Channels' and, in comments on Lopburi, he says 'The King of Siam does there spend the greatest part of the year, the more commodiously to enjoy the diversion of Hunting; but Louvo [Lopburi] would not be habitable, were it not for a channel cut from the River to water it.'[34]

Canal construction continued into the eighteenth century. In 1704–5, King Phra Phuttha Chao Sua (r. 1703–9) began excavating the first major canal in recorded history—the Khlong Khok Kham (later renamed Khlong Mahachai) to connect the Chao Phya and Tha Chin Rivers. Originally dug for transport purposes, the *khlong chuam maenam* would play an important role in enabling Siamese troops to repel Burmese invaders after Bangkok was established as the capital in 1782. Begun in 1705 by Phra Sampetch, it was dug in two parts. The first, called Khlong Phra Phuttha Chao Luang, extended from the Chao Phya River to Khlong Khok Kham. The second section, called Mahachai Cholamak, ran 13.6 kilometres from there to the Tha Chin River. Completed in 1705 by 30,000 conscripted Thai labourers, the 16-metre-wide, 3-metre-deep canal provided easy passage from the Chao Phya to the mouth of the Tha Chin at the Gulf of Thailand.

LIQUID ROAD TO RICHES

It is noteworthy that a foreign engineer was employed to survey the entire route to ensure that it ran straight and true. The Phra Paramanuchit version of the Royal Chronicle of the Ayutthaya period also notes that in inaugurating the work, Phra Phuttha Chao Sua praised the canal excavation works initiated by previous kings, suggesting an earlier, unrecorded history of canal engineering in Ayutthaya.[35]

Fortifications on the Lower River

Whereas early river works were designed to facilitate travel, later projects were attempts to hinder passage. Fortifications and watch-towers were built along the river-bank from Ayutthaya to the sea to monitor ship movements up and down the Chao Phya. The key settlement by the seventeenth century was Thonburi. Founded in 1563, it was called Thonburi Si Mahasamut (Oceanic City of Great Wealth) by the Siamese; it is noted on European navigation charts as any one of several variant spellings of 'Bankok'. While Thonburi was ostensibly a customs port, Gervaise saw its importance in military terms: 'Bangkok is assuredly the most important place in the kingdom of Siam, for it is the only place anywhere on the sea coast that could offer some resistance to enemy attack.'[36]

To gain a picture of life along the river at this juncture, it is useful to read the journal of Engelbert Kaempfer, a Dutch visitor who paused briefly at Ayutthaya on his way to Japan in 1690.

> The mouth of the Meinam opens itself into the sea, as it were, between two wings of low marshy land.... Not far off we saw some batteries planted with cannons on both sides of the River.... About noon we arrived safely at the Dutch habitation and storehouse called Amsterdam [or Nieu Amsterdam] near two leagues distance from the mouth of the River.... I tried to walk about in the adjacent woods which is dry, infested with tigers and other voracious beasts.... The banks of the river are low, and for the greater part marshy ... [and] from Bangkok to the harbor there is nothing but forest.
>
> [Beyond this point] three sorts of animals afford much diversion to travelers sailing on the river. First are to be seen incredible numbers of monkeys of a blackish color.... The glow worms represent another show, which settle on some trees, like a fiery cloud, with this surprising circumstance, that a whole swarm of these insects, having taken possession of one tree and spread themselves over its branches, sometimes hide their light all at once, and a moment after make it appear again with the utmost regularity, as if they were a perpetual systole and diastole. [Numerous other writers were awed by this natural phenomenon.]
>
> The banks above Bangkok are pretty well inhabited and flocked with villages, the houses of which are raised on piles, but built of poor light stuff, and sometimes fine temples and habitations for the priests, with an abundance of trees some bearing fruit, some not.[37]

By the third quarter of the seventeenth century, foreigners had become so well established in Ayutthaya that a Greek, Constant

Phaulkon, was serving as Prime Minister to King Narai. Many Thai nobles perceived this as foreign meddling in their internal affairs and when King Narai died in 1688, they expelled many of the European merchants and severely curtailed the activities of the remainder for the next 140 years. Until the 1820s, the Siamese would concentrate on trade with China and with Asian neighbours.

Ayutthaya's Fall

By the 1760s, internal dissension had weakened the Ayutthayan court and when the Burmese launched a new invasion, the Siamese were unprepared to repel it. The Burmese had also learned how to overcome the watery barriers that had protected Ayutthaya. With artillery, they could breach the walls or lob shells over them to destroy the city's defenders. With boats of their own, they could cross the water and storm the walls. In 1766, they laid siege to the city. Hard-pressed, Siamese King Suriyamarin consigned his kingdom's destiny to the skies and the rivers, hoping that once again, the monsoon rains and floods would force the Burmese to retreat. This time, however, the Burmese responded by fortifying their positions and moving their troops by boat which they commandeered from the populace. The governors of Phetchaburi and Tak led water-borne counter-attacks but the Burmese repulsed them. In the end, the river could not save Ayutthaya. Relentless pressure, famine, epidemics, and finally a fire which destroyed 10,000 houses within the city walls weakened the Siamese beyond retaliation. On 7 April 1767, the Burmese breached the walls, pillaging the city and then torching it. One of their first acts was to destroy the warships which had so successfully harassed and repelled them.

So thorough were the Burmese in slaying, enslaving, or scattering its population, that for decades afterwards, Ayutthaya was a ghost town inhabited by scavengers who eked out an existence amidst the debris of the once-glorious city. Thus did Ayutthaya's role as Siam's capital die in the fires that consumed it, ironically surrounded by a ring of water which could have extinguished the flames.

---

1. Wilaiwan Khanittanan, 'The Order of the Natural World as Recorded in Tai Languages', in *Culture and Environment in Thailand: A Symposium Sponsored by the Siam Society*, Bangkok: Siam Society, 1989, p. 237.

2. Nid H. Shiranan, 'The Contemporary Thai City as an Environmental Adaptation', in *Culture and Environment in Thailand: A Symposium Sponsored by the Siam Society*, Bangkok: Siam Society, 1989, p. 374.

3. *Ancient Customary Laws and History Laws of the North*, Microfilms of palm leaf manuscripts dating from Mengrai's reign, Chiang Mai University library.

4. W. J. Van Liere, 'Mon–Khmer Approaches to the Environment', in *Culture*

and Environment in Thailand: A Symposium Sponsored by the Siam Society, Bangkok: Siam Society, 1989, p. 157.

5. Jonathan Rigg, 'The Gift of Water', in Jonathan Rigg (ed.), *The Gift of Water: Water Management, Cosmology and the State in South East Asia*, London: School of Oriental and African Studies, University of London, 1992, p. 5.

6. Yoneo Ishii, 'History and Rice-growing', in Yoneo Ishii (ed.), *Thailand: A Rice-growing Society*, trans. Peter Hawkes and Stephanie Hawkes, Honolulu: University Press of Hawaii, 1978, p. 25.

7. Ibid., p. 25.

8. Thiva Supajanya, 'Sukhothai: Its Hydraulic Past', *Geology*, Bangkok, Vol. 15, No. 1, 1992, p. 8.

9. Ibid., p. 4.

10. Winai Pongsripian, 'A Historian's Comment', *Geology*, Bangkok, Vol. 15, No. 1, 1992, p. 10.

11. David K. Wyatt, *Thailand: A Short History*, Bangkok: Yale University Press/Thai Wattana Panich, 1984, p. 65.

12. Sumet Jumsai, *Naga: Cultural Origins in Siam and the West Pacific*, Singapore: Oxford University Press, 1988, p. 161.

13. Wyatt, *Thailand: A Short History*, p. 69.

14. Ibid., p. 86.

15. 'Burmese Invasions of Siam, Translated from the Hmannan Yazawin Dawgyi', in Nai Thien (trans.), *Selected Articles from the Siam Society Journal, Relationship with Burma—Part 1*, Bangkok: Siam Society, 1959, p. 8.

16. Samuel Purchas, *Purchas: His Pilgrimage or Relations of the World and the Religions*, 3rd edn., London, 1617, p. 558.

17. Nicolas Gervaise, *The Natural and Political History of the Kingdom of Siam*, Paris: Claude Barbin, 1688; English edn., trans. John Villiers, Bangkok: White Lotus, 1989, p. 13.

18. Ibid., p. 113.

19. Simon de la Loubère, *A New Historical Relation of the Kingdom of Siam*, London, 1693; reprinted Kuala Lumpur and Singapore: Oxford University Press, 1969 and 1986, p. 4.

20. Ibid., p. 6.

21. Sumet, *Naga*, pp. 162–3.

22. Shiranan, 'The Contemporary Thai City as an Environmental Adaptation', p. 300.

23. Gervaise, *The Natural and Political History of the Kingdom of Siam*, p. 50.

24. La Loubère, *A New Historical Relation of the Kingdom of Siam*, p. 4.

25. Ibid., p. 3.

26. Ibid.

27. Ibid., p. 4.

28. Gervaise, *The Natural and Political History of the Kingdom of Siam*, p. 14.

29. La Loubère, *A New Historical Relation of the Kingdom of Siam*, pp. 3–4.

30. Gervaise, *The Natural and Political History of the Kingdom of Siam*, pp. 13–14.

31. La Loubère, *A New Historical Relation of the Kingdom of Siam*, p. 41.

32. Gervaise, *The Natural and Political History of the Kingdom of Siam*, p. 115.

33. All information relating to canals is from: Piyanart Bunnag, Duangporn Nopkhun, and Suwattana Thadaniti, *Khlong Nai Krungthep* [Canals in Bangkok], Bangkok: Chulalongkorn University, 1982.

34. La Loubère, *A New Historical Relation of the Kingdom of Siam*, pp. 4–6.

35. *Phra Ratcha Phongsawadan Krung Si Ayutthaya chabap Phra Paramanuchit* [The Phra Paramanuchit Version of the Royal Chronicles of the Ayutthaya Dynasty], Bangkok: Ongkankha Khong Khrusapha, 1961, Vol. I, pp. 204–5.

36. Gervaise, *The Natural and Political History of the Kingdom of Siam*, p. 49.

37. Engelbert Kaempfer, *A Description of the Kingdom of Siam 1690*, reprinted Bangkok: White Orchid Press, 1987, pp. 19–20, 78–9.

3   Link to the World

Meinam in the Siamite language signifies mother of humidities, which name hath been given to this river by reason of the abundance of its water, which renders the whole country fruitful.[1]

DESPITE the Chao Phya's failure to protect Ayutthaya, Thais continued to adhere to riverine modes of life long after the city's destruction. Thus, rather than establishing a hilltop fortress in a mountainous region like Phetchabun, which would have put additional distance between themselves and the Burmese, the Thais chose yet another river-bank for their new capital city, and a most unlikely site at that. Moving down the Chao Phya, they settled at Thonburi on marshy land subject to floods, protected only by a low wall, and even more directly in the path of the traditional Burmese invasion route down the Mae Klong River Valley. Moreover, the new capital was only partially protected against sea-borne attack by Pom Wichaprasitwong, a fortress built in 1557 at Thonburi's south-eastern corner, at the entrance to Khlong Bangkok Yai.

The selection of Thonburi was more likely one of expedience than of considered choice. King Taksin needed an established base from which to repel the Burmese and to carry on trade to finance his wars. Thonburi had secure moorages, wharves, warehouses, Chinese merchants conversant with mercantile trade (Taksin claimed Chinese ancestry and could rely on their financial aid), and good access to oceanic trading routes. This combination of factors made the town an ideal site from which to speed the reconstruction of his shattered kingdom and its economy.

Amidst the demands of repelling continual invasions, Taksin carried on the river engineering work of his predecessors. In 1771, he dredged a silted Ayutthaya-period canal along Thonburi's western perimeter to encircle the area with a protective moat. Called Khlong Rop Muang (Canal Around the City), it linked Khlong Bangkok Noi on the north with Khlong Bangkok Yai on the south, and made an island of the town. There is some evidence that Taksin planned to construct another palace on the eastern bank because although it was more prone to flooding, it offered more space for future growth than the *wat-* and market-crowded western bank. For that reason, in the same year, he excavated Khlong Lawd 350 metres east of the river, thereby transforming Bangkok[2] into an island (Map 8). Taksin's energies and resources were so fully utilized in

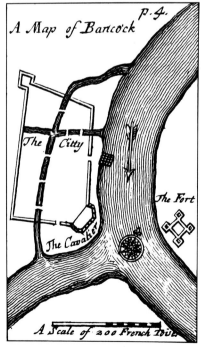

Map 8. An Ayutthaya-period map of 'Bancock'. (la Loubère, 1693, facing p. 7)

defending his kingdom, that he had scant time to build his Thonburi palace much less erect a glorious city on the eastern bank. That was left to his successor, General Chakri, who established his own dynasty after Taksin was declared insane, deposed, and executed in 1782.

It has been suggested that in building his royal city Rama I—the monarchical name Chakri chose for himself—was guided by a desire to re-create the physical structure of Ayutthaya and thereby appropriate its glory and, by extension, the mantle of authority of its fabled monarchs. While he undoubtedly understood the symbolic significance of these acts, he was also a supreme pragmatist, as is evident in the Dynastic Chronicles entry describing his 1782 decision to transfer the seat of power to the eastern bank of the Chao Phya:

If the capital city were to be established on the east side, it would be easier to defend in the event that enemies approached than if it were on the west bank. The only drawback to the east bank was that it was lowland. That was why the King of Thonburi had chosen the west bank, which was higher. However, the west bank was on the concave side, subject to the constant force of the coursing river which caused the land to erode continually, and therefore it was not durable ground.[3]

For practical as well as symbolic reasons, Rama I dismantled Ayutthaya's walls and floated the large square bricks downriver on barges to construct Bangkok's defensive walls. He also employed the barges to transport many of Ayutthaya's large Buddha images to Bangkok's new *wat*. It must have been a splendid day when a grand river procession accompanying the most celebrated image— an 8-metre-tall, 6-metre-wide bronze statue destined for Wat Suthat—arrived at the landing near the Grand Palace.

In 1783, Rama I widened and deepened Khlong Lawd, and then dug a semicircular canal called Khlong Rop Krung (Encircle the Capital Canal) 800 metres to the east and parallel to Khlong Lawd (Plate 14). The latter canal, 4.45 kilometres long, comprised two sections: a 3-kilometre northern portion running between Wat Len and Wat Banglamphu called Khlong Banglamphu, and a 1.5-kilometre second section from Wat Saket to the river at Pak Khlong Talat, a portion later named Khlong Ong Ang (Water Jar Canal) for the glazed jars produced and sold on its banks. The entire project was carried out by 10,000 conscripted Khmer war prisoners who laboured two years to dig the 20.0-metre-wide, 2.5-metre-deep waterway. Along its inner rim, a 4-metre tall crenelated brick and stucco wall was built; at 350-metre intervals, 14 watch-towers mounted with small cannons were erected.

The Dynastic Chronicles also reveal a whimsical side to Rama I:

Another large canal [running east to the present Pratunam, i.e., Watergate, at Ratchadamri Road] was also constructed just north of the Sakae Temple [soon after renamed Wat Saket]. This canal was named the Mahanak [Great Serpent] Canal [Colour Plate 6] by the king. He wanted it to be a place where the people of the capital city could go boating and singing and

14. Khlong Lawd defined the eastern boundary of the royal enclave in the oldest section of Bangkok.

reciting poems during the high-water season, just like the custom observed in the former capital at Ayutthaya.[4]

Together, the river and the new, wall-rimmed canals formed a watertight protective seal around the city. Rama I and his brother, Prince Surasihanat the heir apparent, wanted to construct a bridge across Khlong Rop Krung just south of the mouth of Khlong Mahanak to allow elephants to pass through the city gates. The Abbot of Wat Photharam (Wat Pho) advised that such a bridge not only would allow an enemy entry but would obstruct boat processions, an indication that many Ayutthayan rituals had been preserved. Convinced by this logic, the monarch abandoned his bridge plans. One assumes that the city did not remain entirely water-bound but that temporary, easily dismantled bridges were constructed.

Bangkok's proximity to the sea allowed brine to flow upriver through the Pak Lat Canal at Ratburana, killing fish and, when the river overflowed, destroying vegetation in Bangkok. The Dynastic

Chronicles for 1784 note the measures prescribed to remedy the problem:

> On Friday, the eleventh day of the waxing moon of the third month, the king, together with the heir apparent and other members of the royal family who hold the 'krom' [ministerial] rank, went by boat to oversee the damming of the Paklat Canal. Common people were conscripted at the king's order to pile up dirt and brick, making a dam to prevent the flow of water.[5]

In the succeeding decades, the royal city, bounded by the Chao Phya and the Rop Krung Canal, formed the nucleus of Bangkok. In the nineteenth century, the city would grow to the north and south along the Bangkok bank of the river as well as to the east, beyond the city moat into the rural districts where most of the commoners dwelt. Just north of Khlong Lawd, the riverine street of Phra Athit became a royal enclave lined with fine homes belonging to lesser nobles. Thonburi, in the vicinity of Taksin's palace, became an extension of Bangkok with nobles' houses situated on both sides of the town walls. By the reign of King Rama V (1868–1910), the area would hold more than 100 mansions of which but a few remain today. Chinese merchants, who had vacated the eastern bank so the Grand Palace and Wat Phra Kaew could be built, moved downriver to the orchards of Sampheng, erecting homes and shops and lining the waterfront with new wharves and warehouses. Other Chinese families settled on the Thonburi bank just south of the present Memorial Bridge.

The Chronicles note the importance of the river in sea trade and the expansion along the Chao Phya's shores of the shipbuilding industry begun in Ayutthaya in the late 1600s. After listing various cash taxes on which the royal court depended—liquor, gambling, water (levied on fishing equipment), fruit trees—as well as the tax paid in kind on rice (all of which were characterized as paltry), the Chronicles for 1808 record the following:

> The greatest revenues in that era came from the Chinese junk trade.... The junks measured each from five to seven *wa* [10–14 metres] across the beam. There were a large number of them. Some belonged to the Crown, some to members of the royal family, some to government officials, and some to merchants.... These ships were either built in Bangkok or in provincial areas outside the capital. They were loaded with merchandise to be sold in China every year. Some ships sold only their cargoes of merchandise, some ships sold both their cargoes of merchandise and the ships themselves as well. The profits from this junk trade were tremendous.[6]

The River in War

These boatbuilding skills were also employed to create warships. Documents of the period portray Bangkok as a bustling metropolis whose tranquillity was continually plagued by the threat of invasion. The river and its tributaries served as the principal channels along which to speed troops to defend the frontiers. The Chronicles repeatedly refer to armies taking the 'water route' out of the city and some of the departing processions sound quite splendid. An account of a battle with Burmese forces at Kanchanaburi in 1785 notes that

> His [the king's younger brother] boat was preceded by the royal barge the Sawatdichingchai, which was painted black and was carrying the Buddhist statue Phra Chai. Other titled royal attendants followed in a large number of boats. The royal army thus departed, proceeding by the water route.[7]

Their journey would have taken them along Khlong Phra Phuttha Chao Luang dug in 1705, then along its continuation, Mahachai Cholamak, to the Tha Chin River at Samut Sakhon. At that point, a natural canal, the Sunak Hon (Barking Dog Canal), wound through the flats to Samut Songkhram on the Mae Klong River, for a hard paddle upriver to Ratchaburi and from there to Kanchanaburi. In the reign of Rama II, the Sunak Hon would be widened, deepened, and straightened.

In a 1785 skirmish against the Burmese, the fleet paddled up the Chao Phya to Inthaburi. The troops continued on foot to fight two major battles on the banks of the Ping, one near Paknampho (Nakhon Sawan) and another at Raheng (Tak). The Chronicles note that the river was so polluted by the corpses of dead Burmese that it was undrinkable, an interesting comment because it has been many years since river water has been considered potable.

Bangkok's proximity to the sea would pose a defence dilemma for its rulers until the beginning of the twentieth century. In the 1780s, Burmese armies had crossed southern Thailand at Kanburi, sailed up the Gulf and into the Chao Phya, capturing the rivermouth town of Samut Prakan (Fortress by the Sea). Rama I recaptured it but feared a repeat invasion, either by the Burmese or by Ong Chiang Su, a Vietnamese prince who had requested asylum in 1782 and subsequently learnt a great deal about Bangkok's fortifications but who fled Thailand in 1786 to establish his claim to the Vietnamese throne. Prince Surasihanat, the heir apparent, confided his fears to his brother, Rama I: 'Right now at Samut Prakan, there is not one thing that could meet a challenge from enemies coming from the sea.... I ask permission to build a town at the estuary.'[8] No town was built but a fortress named Pom Withayakhom was completed before Rama I's death in 1809. In 1814, Rama II built a garrison town at the mouth of the Pak Lat canal. He transferred 3,000 newly arrived Burmese Mon refugees, then residing at Pathumthani, to the new town which he called Nakhon Khuan Khan, later renamed Phra Pradaeng. The Dynastic Chronicles note that these new residents built three forts on the eastern bank, and five on the western. The fortresses would be strengthened and bolstered with cannons in the succeeding reign, to confront an old antagonist, the Europeans.

Given the strategic importance of the Chao Phya, one would think that rulers would have assigned foreign refugees to areas as far from the rivers as possible. Yet, the first three reigns saw the growth of numerous towns inhabited by single nationalities. In 1782, Cambodian Catholics built houses just north of Wat Samorai, later renamed Wat Rachatiwat, the future site of the National Library. Lao, Khmer, and Mon war captives responsible for building walls and canals were clustered in communities along the river and canals. In the 1830s, Vietnamese Catholics fleeing persecution were settled in what would be known as Ban Yuan (Vietnamese Village) just south of the present Krung Thon Bridge. Later, up-river towns would be inhabited entirely by Chinese immigrants, giving the streets and buildings a distinctly Sino-Thai flavour.

Canal construction slowed during the reign of Rama II as the focus shifted to widening and deepening several old canals. One major new canal, Khlong Lat Lang Muang Nakhon Khuan Khan was dug south of Bangkok to connect Nakhon Khuan Khan (i.e. Pak Lat Phra Padaeng) with Khlong Ta-lao. The canal facilitated communications between the capital and the coastal areas which formed a barrier against invasion from the sea. Its mode of excavation marked a departure from traditional practice. Canals had formerly been dug by foreign war prisoners or by Thais engaged as part of their annual corvée labour requirements. As hostilities with neighbours had abated, the supply of war captives had dwindled. Moreover, recognizing that Thai farmers had insufficient time for these duties, Rama III (r. 1824–51) hired Chinese immigrants then pouring into Thailand, paying them from the privy purse, an approach followed in the ensuing decades.

The Return of the West

Security considerations continued to be the dominant rationale for canal construction well into Rama III's reign. An extension of Khlong Mahanak beginning at Pratunam ran slightly north-east for 33.5 kilometres to the Bangpakong River. Between 1837 and 1840, Chinese labourers dug the 12-metre-wide, 2-metre-deep canal to speed troops to defend—and Bangkok officials to administrate—the eastern provinces. A fleet of several hundred naval barges was stationed at Hua Mak in specially built sheds. The canal was also heavily trafficked by barges carrying rice to Bangkok's markets or to its waterfront where it was loaded on ships bound for foreign ports. Like many canals, the extension bore two names; the Bangkok end was known as Saensap and the Bangpakong portion was called Bang Kanak.[9]

By the 1830s, the European colonialist threat was being felt in the Thai royal court. In 1826, Rama III had concluded the Burney Treaty of Amity with the British, and in 1833, he signed a Treaty of Commerce and Amity with the United States, opening the doors to foreign commercial activity which had been closed since 1688. A trickle and then a flood of foreign vessels began calling at Bangkok. As his reign progressed, Rama III became increasingly xenophobic, distrustful of foreigners' motives, and paranoid about the possibility of armed confrontations. As a result, he felt compelled to bolster the city's downriver defences. Nakharat and Tapsamut Fortresses were built in Samut Prakan province and numerous cannons, engraved with the names of their Chinese, British, and American manufacturers, were installed in them. The Dynastic Chronicles of the Second Reign (1809–24) describe attempts to stretch a chain across the river as a barrier which could be raised and lowered by pulleys to halt the passage of ships. The Chronicles note that the brick supports failed to hold it securely to the banks even after reinforcement. Accordingly, in 1826, at Samut Prakan, a second chain system was constructed. To ensure that an enemy ship would stay within range of shore batteries, several old junks were loaded with stones and sunk in the channel. Upstream, the Siamese anchored a number of rafts laden with firewood. Were an enemy ship to breach the chain and evade the cannons, Siamese soldiers would ignite the rafts and allow them to drift downstream into the invaders.

In the same year, a small fortress bristling with cannons was constructed on a mid-river island. Adjacent to the cannons, the Buddhist Phra Samut Chedi (Ocean Stupa), popularly known as Wat Phra Chedi Klang Nam (Stupa in the Middle of the River), was erected, ultimately acting as a beacon to travellers returning home from abroad (Colour Plate 7). Beginning in the 1880s, when King Chulalongkorn sailed across the seas to and from state visits, royal passengers would disembark at the *wat* to pay obeisance. In the 1900s, the river would shift its course to the east and the right channel would eventually silt up, anchoring Phra Chedi Klang Nam to the western bank. Until the 1950s, ships continued to pause at the *wat*, but soon after, airplanes superseded ocean liners as the preferred mode of royal transport, and the ritual ceased.

With the revival of trade with Europe, the river regained its former importance. Henceforth, canals would be dug to serve commerce, transport goods, or facilitate provincial administration and tax collection. By 1825, at the time of John Crawfurd's visit, the Chao Phya was the stage for a lively trade in the goods of the world (Map 9).

> The Menam exhibited a scene of considerable activity, and afforded evidence of the existence of a respectable trade. We counted seventy junks, large and small, engaged in foreign trade.... Besides these, there [were] numerous small craft engaged in the internal trade, as well as many rafts conveying merchandize [Plate 15].[10]

Crawfurd commented on the enormous variety of produce which flowed down the canals and rivers and into the holds of ships bound for foreign ports (Plate 16). His list suggests the breadth of

▷ Map 9. Crawfurd's map of the river displays foreigners' confusion about the upper portion of the Ping River. (Crawfurd, 1828, facing p. 1)

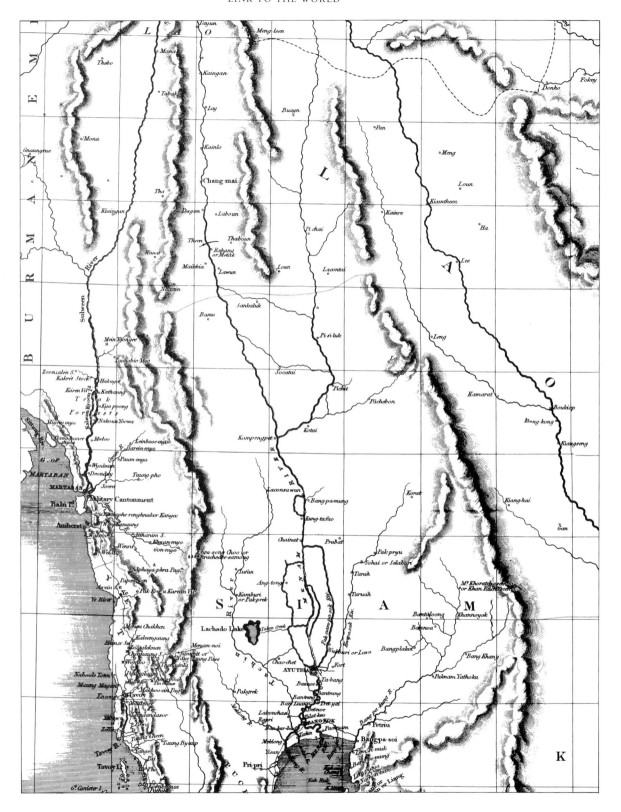

15. Lighters and other boats on the Chao Phya below Bangkok. (Courtesy National Archives)

16. Congestion on Khlong Ong Ang in the late nineteenth century, framed by the Saphan Han bridge.

the trade and the size of the fleet required to transport the products from deep in the forests to Bangkok:

> ... black pepper, sugar, tin, cardamums, eagle-wood, sapan-wood, red mangrove bark, rose-wood for furniture and cabinet work, cotton, ivory, stic-lac, rice, areca-nuts, salt-fish, the hides and skins of oxen, buffaloes, elephants, rhinoceros, deer, tigers, leopards, otters, civet cats, the pangolin; of snakes and rays, with the belly-shell of a species of land tortoise; the horns of the buffalo, ox, deer, and rhinoceros; the bones of the ox, buffalo, elephant, rhinoceros, and tiger; dried deers' sinews, the feathers of the pelican, of several species of stork, of the peacock and kingfisher, &c. and finally esculent swallows' nests.

Elsewhere, he writes that the banks of Khlong Bangkok Yai, across the river, were crowded with floating houses and shops selling domestic commodities which suggested a very modest diet and humble needs: 'The canal was crowded with merchant-boats, loaded with rice, salt, cotton, dried fish, oil, dyewoods....'[11]

Trade spurred the excavation of new canals. In the same year that he ascended to the throne, King Rama IV (r. 1851–68) hired Chinese workmen to dig a 3.42-kilometre canal, Khlong Padung Krung Kasem. Measuring 20 metres wide and 3 metres deep and extending from Wat Kaew Chan Fa to Wat Thewaratchakunchan and in an arc beyond but parallel to Khlong Lawd and Khlong Banglamphu/Ong Ang, this was the third and last of the concentric metropolitan canals (Map 10). At the request of foreign traders who wanted a shorter route from their godowns in Phrakhanong east of Bangkok to the city centre, a canal was dug from Bangna

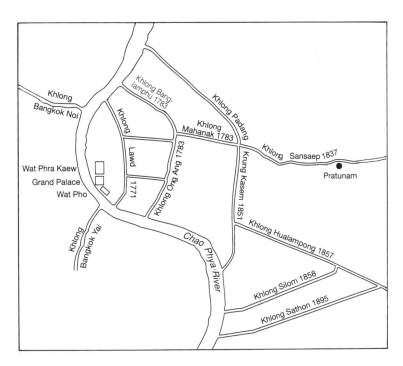

Map 10. Bangkok canals dug during the past two centuries.

17. The canalside offices of one of Bangkok's oldest European trading companies in the 1920s. (Private collection)

to intersect with Khlong Padung Krung Kasem at what is now the Hualampong Railway Station. Completed in 1857, the new canal, called Wat Wua Lampong (later changed to Hualampong), was known familiarly as Thanon Trong (Straight Street). At 5.18 kilometres long, 12 metres wide, and 3 metres deep, it eliminated a 20-kilometre journey along the bends of the river. The excavated soil was heaped on both canal banks to provide a bed for the street later known as Rama IV Road. Similarly, Khlong Silom (Windmill Canal) was financed by the privy purse for transporting cattle to the markets at Saladaeng near the present intersection with Rama IV Road. Nearly 1.7 kilometres long, 12 metres wide, and 3 metres deep, the canal was completed in 1861.

The canals and roads soon became foci of urban development, serving both commerce and daily life. As Bangkok neared the end of its first century, mercantile firms erected offices on canal and river-banks where they conducted the principal trading between the city and the countryside, and between Bangkok and overseas partners (Plate 17). As Thailand entered the twentieth century, land and rail transport was regarded as speedier than water conveyance. The warehouses became mere cargo transfer points and offices moved deeper into the city. Ultimately, the canals would be abandoned altogether, filled in to create roads so trucks could travel swiftly to distant markets.

Floating Communities

A prominent feature of Bangkok's waterfront throughout the nineteenth century was its large community of floating houses. Foreign accounts reveal a great deal about life in these water-borne villages. The first to describe them in print was Engelbert Kaempfer. Writing in 1690 about Ayutthaya and its environs, he noted:

Round the city lie many suburbs and villages some of which consist of inhabited ships or vessels rather than houses containing two, three, or more families each; they remove them from time to time and float them particularly when the waters are high, where fairs are kept to sell their goods there and get their livelihood by it.[12]

By the 1800s, the Chao Phya at Bangkok was both a town and a thriving bazaar. Crawfurd described this lively scene on first arriving in Bangkok:

On each side of the river there was a row of floating habitations, resting on rafts of bamboos, moored to the shore ... occupied by good Chinese shops.... The number of these struck us as very great at the time, for we were not aware that there are few or no roads at Bang-kok, and that the river and canals form the common highways, not only for goods, but for passengers of every description. Many of the boats were shops containing earthenware, blachang [fish sauce, a condiment], dried fish, and fresh pork. Vendors of these several commodities were hawking and crying them as in an European town. Among those who plied on the river, there was a large proportion of women, and of the priests of Buddha; the latter readily distinguished by their shaved and bare heads, and their yellow vestments. This was the hour in which they are accustomed to go in quest of alms, which accounted for the great number of them which we saw.[13]

By the 1830s, with riverine commerce even more frenetic, the line of demarcation between the river and land was barely discernible. W. S. W. Ruschenberger reported:

On each side of the river are moored rafts of bamboos on which are constructed houses or sheds with open verandas in front, wherein various goods are exposed for sale. Rows of Chinese junks, some of them several hundred tons, extend two miles along the middle of the stream, retailing their cargoes. Many families live wholly in their little gondolas, called sampans.[14]

In 1857, rapturous foreign enthusiasm for this floating town prompted a wry comment by King Mongkut in a letter to his ambassador in London:

A great number of Englishmen have been and are now residing in this country. They seem to have an accurate knowledge of everything that is to be known here, but it is rather regrettable that they still retain a fixed idea regarding four phenomena characteristic of this country, [one being that] three-quarters of the houses in Bangkok are built in the water, only one quarter being built on dry land.[15]

Historian B. J. Terwiel notes that nearly all estimates of the number of houseboats were wildly exaggerated. Quoting a typical figure, that provided by F. A. Neale, who, in his 1852 book

Narrative of a Residence in Siam, claimed that there were 70,000 floating houses, Terwiel comments that

> 70,000 houseboats would cover a shoreline of no less than seventy kilometres. It is well known from our travellers' accounts that such floating houses were found on both sides of the Chao Phya River over a few kilometres of shore. Therefore a few thousand houseboats may already be a generous estimate for Bangkok.[16]

Upriver Commerce after 1880

From contemporary European accounts, the river-banks above Bangkok sound idyllic, with towns set like islands in seas of primary forest. Even the river-banks a short distance north of Bangkok were still covered in trees:

> As we approach Ayuthia there are many monkeys in the trees; big and little ones come down to the water and play along the shore, among the exposed roots of trees, so quietly and so naturally that one is almost tempted to imagine them to be a diminutive hairy race of human beings.... Parrots,

18. Working a merchant boat upriver through the Kaeng Soi Rapids by hauling on long lines. (Private collection)

which are known only as prisoners in America, fly hither and thither on the banks of the Meinam.[17]

Land trade between Bangkok and the northern kingdoms, and beyond them to southern China, was augmented by riverine commerce that made riparian villages, notably those at river confluences, into important trade centres. By the 1890s, the rivers were the principal conduits for logging operations in the North, for the transport of rice and agricultural goods from north to south, and for retail goods sold in upriver towns (Plate 18). The Royal State Railways statistics for 1903/4 reveal that 97.0 per cent of all rice exports had travelled from the provinces to Bangkok by boat, as had 93.5 per cent of all other types of exports.[18] Other commodities, like sand, pottery, and salt, also reached the capital in the holds of river barges.

Significant for its role in carrying Bangkok culture into the interior and in fostering links between the city and the hinterlands was the retail trade conducted by *rua tung* or 'storeboats'. During the 1870s, Bangkok's trade nodes had been located along its waterfront with retail shops and godowns serving as interfaces between

produce- and product-laden boats and the inner city. The 1880s saw a trade shift from Sampheng to New Road, reducing the role played by small Chinese entrepreneurs. Seeking new outlets for their merchandise, these shopowners looked upriver for trade opportunities, outfitting storeboats powered by newly introduced steam engines to carry goods into the interior.

The two-storey *rua tung* carried goods on the lower decks and passengers on the upper. Both decks were protected from the elements by a roof and by canvas walls which could be rolled up or lashed down. The storeboat merchant dealt in basic consumer goods: canned foods, dried salt-water fish, live animals, fuel oil, ice, whiskey, patent medicines, vegetables, sundries, cloth and clothing, and some small luxury items. As time passed, merchants used their city connections to offer moneylending, rice broking, business middlemen, wholesale distribution, and transport broking services.[19]

The storeboat plied a regular route so his customers could anticipate his arrival. He would remain at a town for up to one week before moving on to the next stop. A journey would normally be planned as a circuit which would bring him back to his starting point after 20–5 days, and its distance could vary between 100 and 190 kilometres.[20] The storeboat network was supplemented by a fleet of smaller boats that concentrated on high turnover perishables. Their shorter circuits enabled the merchants to return to base and replenish their supplies every 7–10 days. Other Chinese shopkeepers built stores next to upriver piers to distribute goods to interior villages. In many instances, a shop grew into a town with a distinctly Chinese atmosphere that prevails even today.

The Rise of the Timber Industry

The late nineteenth century saw the start of another undertaking which would ultimately have a profound impact on Thailand's rivers: teak exports. Crawfurd in 1825 had seen the potential for the trade: 'The Teak forests of Siam appear, by all accounts, to be considerable ... [and] will, in all probability, become a valuable article of trade in our intercourse with the Siamese.'[21] It would be several decades before such hopes came to fruition.

Having exhausted many of the timber resources in Burma, British firms in the 1880s obtained concessions to cut teak in northern Thailand. Over the next several decades, logging would destroy watersheds and remove much of the ground cover which had inhibited erosion. It would also encourage settlement and cultivation which would further exacerbate the release of sediment into what had previously been a clear river.

Numerous British accounts (Reginald Campbell's *Teak Wallah* among them) describe the armies of trained elephants which extracted trees from forests stalked by tigers and other wild beasts. Nowhere in the accounts of timber wealth is there any perception of a finite supply of trees; the forests seemed to stretch far beyond the horizon and imagining. The rivers themselves, notably the Wang and the Yom, served as natural flumes for transporting the logs downriver. Once the logs reached the river-banks, they were rolled into the waterway (Plate 19) and left to make their own way downstream, a process called *ploi khwai* (release the buffaloes). With the annual rise and fall of the river, logs would become stranded on the river-banks or stuck in massive jams that only dynamite could clear. It is estimated that a log cut high up the river could take five years to reach the marshalling points. Lakhon (Lampang) on the Wang, Raheng on the Ping, Phrae and Sawankhalok on the Yom, and Paknampho on the Chao Phya owe their prosperity to their role as assembly points where the free-floating logs were bound into rafts for the journey downstream to Bangkok. The beautiful old homes (Plate 20) along Lampang's Talat Kao Road and many of northern Thailand's finest Burmese-style Buddhist *wat* were financed by timber revenues.

At these towns, the logs were bundled into two types of rafts. If the river was narrow and winding, they were tied in a *phae mai sak bap ngu sai* (teak raft like a slithering snake), which was wider at the back than at the front. If the river was wide, the logs were gathered into a *phae mai sak pla taphian* (teak raft like a taphian fish), which was broad in the middle and narrower at the ends. As a raft rounded a bend, an anchorman would secure a line to the

19. Elephants muscle logs into the Yom River in the forests of the North at the beginning of the century. (Courtesy National Archives)

THE CHAO PHYA

20. One of the many grand old houses on Lampang's river-banks that were built by timber barons.

21. Rafts of bamboo being readied for the long journey down the Ping to Bangkok.

bank so that the raft would swing in an arc to avoid colliding with the opposite shore. Eventually, the raft reached the sawmills on the northern outskirts of Bangkok in the vicinity of the present Rama VI Bridge. Highways sounded the death knell for the *ploi khwai* method of transporting logs. Trucks now carry the logs to a point below Nakhon Sawan where they are assembled into rafts. Similarly, bamboo rafts (Plate 21) are floated to Pak Kret where the bamboo is sorted and sold.

New Perspectives, 1900

With the arrival of the twentieth century, the Chao Phya lost its pre-eminent role in Thai commerce and life. Before enumerating the changes, it is useful at that eleventh hour to view, through

European eyes, the river's importance in daily life. In 1897, Maxwell Sommerville wrote:

Mounting the stream, with the tide against us during several hours, hundreds of houseboats continue to shut out the view, crowded one against and behind another on either side of the Meinam.... The spires of wats on land, the bustling traffic in boats, peddlers, ever alert, sculling from one side of our ship to the other, seeking customers, mendicants, musicians, fishermen, priests—all this extended scene causes us to wonder when we will ever get beyond Bangkok.[22]

In similar vein, Ernest Young observed in 1898:

It is an amusing and not uncommon sight to see a father and his family, aided by a few muscular friends or relatives, tugging away at ponderous shovel-shaped oars, fastened fore and aft, as they pilot their house through a crowd of smaller crafts on their way to settle in some more desirable or convenient locality.[23]

Sommerville also recorded that even at night the river was alive:

The floating theatres add their glare and glitter; the supernumeraries stand well out on the platform, beckoning with their firebrands; others guard the lights of the many-colored paper lanterns; and here is a show where some of the actors stand without, giving tempting examples of the entertainment to be enjoyed within ... on a floating platform in the subdued light, stands a screen, on which a light from within and behind casts a series of silhouettes. The performers' hands and arms are posed in such a manner as to produce representations of birds, animals, and human beings; many amusing contests between the characters and ludicrous predicaments of mirthsome Judys as are thrown on the screen.... The artistic performance of the silhouette-maker is interrupted at intervals by a company of quasi-musicians who, by a terrific blast of horns and ringing of gongs, call on the innocents to try Dame Fortune. These are floating gambling shops, where, at games similar to roulette, so much coin changes hands of an evening that an implement resembling a coal-scoop is used to shovel in the money.... Even at this distance in the bright light, we can see many colors and tinsel moving on the outer stage of the Lakons' theatre. These girls are entertaining and cheering the weary laborers, bringing a veil of mirth between their audience and the troubles of the day.[24]

Jacob T. Child, a contemporary, observed that these floating villages and markets began functioning well before dawn:

It is these floating quarters of the city that hold a market in the middle of the night. Boats assemble from every direction, laden to the water's edge with fruit, vegetables, rice and poultry. Here it is the women who do the business, and by the light of a little lamp made entirely of coconut and burning its oil, the bargaining goes as briskly as on any market day at home. The whole river is covered with these little flickering flames; they gleam everywhere like glowworms and the boats lie so closely that it would be useless to try to make one's way in a steam launch among them. But with the first ray of the sun all these specks of light are extinguished, and where but lately a confusion of cranky craft and yelling humanity might be seen, the broad river lies empty and silent in its quiet everyday mood.[25]

Yet amidst these lively urban scenes, the tenor of Thai life was changing. When King Mongkut built the first city road in 1857 (Rama IV Road) and the first paved road in 1863 (Charoenkrung or New Road) in response to calls for modernization, he initiated a movement away from the river, eschewing boats in favour of horses and carriages. The process continued for the remainder of his reign and through that of King Chulalongkorn as more roads were built and a host of bridges began to span the canals. In 1893, an electric train had been built to connect Bangkok with Paknam, eliminating the need for a boat journey. By 1900, horse-drawn carriages were being replaced by cars, which further reduced the role of transport played by canals.

In the year 1900, King Chulalongkorn signalled even further movement inland by constructing a 'summer' palace, Vimarn Mek, north of the city and 2 kilometres east of the river. Nobles ensconced on Phra Athit Road and in Thonburi vied with one another to erect the grandest new mansions in the vicinity of Vimarn Mek. In so doing, they exchanged a riverine setting, annual floods, boat transport, and cool breezes, for a land-based existence with gardens and reduced flooding, and which favoured the motor car and the roads it travelled. It seems a subtle change but it had considerable impact on the way leaders henceforth viewed the river's role in Thailand's development.

Urban canal construction slowed, then ceased. Khlong Sathorn, excavated in 1895 as the only major new canal in Bangkok at the end of the nineteenth century, extended from the Chao Phya River to Khlong Hualampong at its intersection with Wittayu (Wireless) Road. After 1900, a few Thonburi canals were dredged for transport purposes but for the most part, the age of the urban canal was over; after 1915, no new ones were excavated. Instead, after World War II, city planners began filling them in, transforming them into roads that would speed the delivery of goods to an increasingly impatient populace. In 1955, the last major canal, Khlong Hualampong (Thanon Trong), was buried to widen Rama IV Road. By the 1970s, only the main canals—Khlong Lawd, Ong Ang, Banglamphu, Padung Krung Kasem, Mahanak, Saensap, Hualampong, Samsen, Bang Sue, Promprachakorn, Bang Khen, Prapha, Bangkok Yai, Bangkok Noi, and Mon—remained intact, and were used primarily for drainage. As residents extended their stilted houses into them, the canals began to narrow and to stagnate as increasing amounts of sewage were dumped into them. On the Thonburi side of the river, one sees the occasional barge carrying construction materials to canalside building sites but boats have essentially ceased to serve large-scale commerce on the canal. Even the *khlong* jars and clay pottery formerly transported by canal now arrive in Bangkok on trucks. Villagers go to market; it no longer comes to them.

Riverine and long-distance canal-based commerce has been even more severely affected by changing technology. In 1922, the northern railway reached Chiang Mai and trains took precedence over boats as the principal north–south mode of passenger and goods conveyance. The rice and salt barges formerly paddled along the rivers now carry construction sand and are towed in long chains of up to twenty barges each. Most of the graceful teak barges have been replaced by more durable steel vessels. Even on the bustling Saensap Canal, the barges that once carried produce and consumer goods between Bangkok and the eastern provinces have disappeared, with the last towboat company ceasing operations in 1967.

The municipal authorities in Bangkok began to phase out houseboat communities in the 1920s. By the 1950s, only the barge colonies survived, and in the late 1970s, legislation was enacted to clear the canals of them as well. Of the dozens of riverine barge communities, two remain: one, south of the Krungthon Bridge (Plate 22); and another at the entrance to Pak Khlong Talat. The numbers of barges are dwindling, however, as timber and repair costs escalate. Houseboat communities survive on the Nan River just above Nakhon Sawan where their owners propagate fish in floating pens. They also line the waterfront at Phitsanulok, but the authorities have begun to remove them on the pretext that they pollute the river.

The storeboat trade has fared little better (Plate 23). In the 1930s, the merchant boat system had reigned supreme with, for example, one family operating 70 storeboats between Chainat and Nakhon Sawan. The development of road transport reduced the family's fleet to 20 in the late 1950s, and 6 by the end of the 1960s, as truck delivery took over the distribution of consumer goods.[26] Today, a handful of storeboats serve settlements along the banks of the Sirikit Dam reservoir but have disappeared elsewhere. One occasionally sees their bleached skeletons and those of ruined barges stranded high on the banks of the Nan River, beached like whales by the flood waters.

22. One of the few remaining barge communities is moored below Bangkok's Krungthon Bridge.

23. One of the few storeboats still plying the lower Chao Phya River.

THE CHAO PHYA

In spite of the decline of the timber trade in the 1930s, Lampang, Phrae, Tak, and other riverside towns continued to prosper, and when provincial administration was restructured in the same decade, most of the towns selected as provincial capitals were situated on rivers. Yet, with the coming of roads in the 1960s, many river towns were literally left in the backwaters, their piers bereft of the boats that had formerly crowded them (Colour Plate 8). Towns built on major highways continued to flourish but those which the roads had bypassed, even by as little as 5 kilometres, lapsed into somnolence.

Perhaps most symbolic of the triumph of road and rail over river was the construction of bridges. Given the rudimentary engineering of the time, it is not surprising that the first bridge crossed, not the broad Chao Phya, but the narrower Ping at Chiang Mai. Built in 1884 by the missionary Dr Marion Cheek, the cantilever bridge (Plate 24) permitted two oxcarts to pass each other comfortably; for twenty-three years, it was the only bridge across any Thai river. The appearance of motor cars and a small steamroller necessitated a stronger, broader bridge; thus, in 1907, the Navarat Bridge was built on the main road into the city. The three-span teak bridge designed by an Italian engineer, Count Roberti, served until 1935 when it suddenly collapsed; it was

24. Dr Cheek's cantilever bridge over the Ping at Chiang Mai. (Courtesy National Archives)

replaced by an iron bridge. In the 1920s, the tracks of the northern railway approximated the course of the river. Where it crossed three tributaries, steel bridges were constructed. In most river towns, cross-river traffic was served by ferries (Plate 25), but by the early 1990s, most of those services had been superseded by concrete bridges.

The first bridge to cross the Chao Phya at Bangkok was the Rama VI Bridge, built in 1926 just north of Kiak Kai to link the eastern and western portions of the southern railway line which ultimately ended in Singapore. Bangkok's first vehicular bridge, Phuttha Yot Fa or Memorial Bridge, was opened on 6 April 1932 to commemorate the sesquicentennial of the establishment of Bangkok as Thailand's capital. The Krung Thep and Krung Thon Bridges followed soon after, the latter built as a drawbridge to permit the passage of large boats. By the late 1970s, its central span had rusted shut and, as large boats no longer called at the city, no attempt was made to free it.

From bridges crossing the rivers, it was but a short step to dams obstructing them. Small diversion dams had been built on tributaries but large dams, first proposed in 1902, did not become a reality until 1957 when the Chao Phya Dam was built. It remains the only dam on the Chao Phya but, more significantly, of all the dams since built on any of the tributaries, it is the only one with navigation locks. The construction of lockless dams on the Ping, Nan, and Wang in the 1960s and 1970s was tacit recognition of the declining importance of the rivers as transport routes. Henceforth, inland navigation along the tributaries would cease. The Chao Phya Dam also changed the economics of rice. A major portion of the rice formerly bought and sold in Nakhon Sawan found a new market-place below the dam in Phayuha Khiri, which is now the principal rice-trading centre.

25. A vehicular ferry on the lower Ping.

Today, the wooden houses along the Bangkok waterfront are overshadowed by high-rise condominiums and hotels. The intimate contact between river and land has been destroyed by the construction of sea walls which both confine the river and partition it off from daily life. Discussion is under way on plans to build a major highway down the left bank, a move which would signal the dismissal of the river as an urban entity. The river still swarms with commuter and ferry boats and the occasional fisherman casts a line but more and more the pleasure boats of the affluent chop and churn the waters.

Each of these progressions has marked yet another psychological as well as physical movement away from the rivers, a rejection which has had dramatic repercussions in the late twentieth century. To appreciate the magnitude of these changes, it is necessary to examine the river's role in daily life.

1. Engelbert Kaempfer, *A Description of the Kingdom of Siam 1690*, reprinted Bangkok: White Orchid Press, 1987, p. 75.

2. The name 'Bangkok' is a foreign corruption of the town's Thai name, Ban Ko (Island Town). Most Ayutthaya-period maps use the name to designate the town of Thonburi.

3. Thadeus Flood and Chadin Flood (eds., trans.), *The Dynastic Chronicles Bangkok Era, the First Reign: Chaophraya Thiphakorawong Edition*, Tokyo: Center for East Asian Cultural Studies, 1978, Vol. 1, p. 1.

4. Ibid., pp. 58–9.

5. Ibid., p. 70.

6. Ibid., pp. 303–4.

7. Ibid., p. 91.

8. Ibid., p. 125.

9. Piyanart Bunnag, Duangporn Nopkhun, and Suwattana Thadaniti, *Khlong Nai Krungthep*, Bangkok: Chulalongkorn University, 1982, Supplement III.

10. John Crawfurd, *Journal of an Embassy to the Courts of Siam and Cochin China*, London, 1828; reprinted Kuala Lumpur and Singapore: Oxford University Press, 1967 and 1987, p. 106.

11. Ibid., pp. 408–9, 117.

12. Kaempfer, *A Description of the Kingdom of Siam*, p. 49.

13. Crawfurd, *Journal of an Embassy to the Courts of Siam and Cochin China*, p. 79.

14. W. S. W. Ruschenberger, *A Voyage around the World including an Embassy to Muscat and Siam 1835, 1836 and 1837*, Philadelphia, 1838; quoted in *Foreign Records of the Bangkok Period up to A.D. 1932*, Bangkok: Office of the Prime Minister, 1982, p. 9.

15. MR Seni Pramoj and MR Kukrit Pramoj, 'Letter to Phya Montri Suriwongse, the King's Ambassador to the Court of Queen Victoria, and Chao Mun Sarapethbhakdi, Vice-Ambassador, 1857', in *A King of Siam Speaks*, Bangkok: Siam Society, 1987, p. 212.

16. B. J. Terwiel, *Through Travellers' Eyes: An Approach to Early Nineteenth Century Thai History*, Bangkok: Editions Duang Kamol, 1989, p. 230.

17. Maxwell Sommerville, *Siam on the Meinam; From the Gulf to Ayuthia*, London: Sampson Low, Marston and Company, 1897; reprinted Bangkok: White Lotus, 1985, p. 107.

18. A. Cecil Carter, 1923, p. 230, as quoted in James A. Hafner, *Salt, Seasons and Sampans: Riverine Trade and Transport in Central Thailand*, Asian Studies

Committee Occasional Papers Series No. 6, International Area Studies Program, University of Massachusetts at Amherst, 1980, p. 11.

19. Hafner, *Salt, Seasons and Sampans*, p. 16.

20. James A. Hafner, 'Riverine Commerce in Thailand: Tradition in Decline', *Journal of the Siam Society*, Vol. 62, Pt. 2, 1974, p. 17.

21. Crawfurd, *Journal of an Embassy to the Courts of Siam and Cochin China*, p. 427.

22. Sommerville, *Siam on the Meinam*, p. 95.

23. Ernest Young, *The Kingdom of the Yellow Robe: Being Sketches of the Domestic and Religious Rites and Ceremonies of the Siamese*, London: Archibald Constable & Co., 1898; reprinted Kuala Lumpur: Oxford University Press, 1982, p. 34.

24. Sommerville, *Siam on the Meinam*, pp. 52–3.

25. Jacob T. Child, *The Pearl of Asia*, quoted in *Foreign Records of the Bangkok Period up to A.D. 1932*, Bangkok: Office of the Prime Minister, 1982, p. 211.

26. Hafner, 'Riverine Commerce in Thailand', p. 18.

4 Adapting Life to the River's Rhythms

The water is the true home of the Siamese, and it is on this, their native element, that their real character and genius are best exhibited.[1]

ERNEST YOUNG'S observation in 1898 captured the essence of the bond between Thais and rivers. Recognizing the futility of contending with the rivers' power, Thais harmonized their lives to conform with riverine rhythms, designing towns and villages, houses, boats, occupations, and lifestyles according to their moods and seasons. In bowing to them, they gained partial ascendancy over them, and were stymied only when the contrariness of the skies resulted in drought or floods, or when water-borne disease ravaged their lives.

House Design and Settlement Patterns

House design and village settlement patterns in the Chao Phya basin reveal much about Thai attitudes towards their rivers and their complex relationships with them. As one descends the rivers from their sources to the sea, one is struck by two synchronous phenomena: houses move from the hilltops to the valleys and then edge closer to the rivers until, by the Central Plains, they overhang the Chao Phya's banks. Simultaneously, the homes grow taller on their stilt legs, as if rising from a squatting to a standing position.

Northern Villages

The headwaters of the four tributaries are inhabited by tribal peoples who look to the skies for sustenance and seem almost dismissive of the rivers. In the past, these nomadic tribes pursued a destructive form of cultivation known as swidden, or slash-and-burn, felling virgin forest, burning off its underbrush, and planting maize, rice, and vegetables which were watered by the rains. When soil fertility was exhausted, they migrated elsewhere to repeat the process, leaving the old area to recover as best it could. Thai government agricultural extension programmes of the 1970s taught the tribesmen how to maintain soil fertility and, in the process, succeeded in persuading them to remain in one location and build permanent homes.

In the transition from a nomadic to a stable existence, tribal house and village design has changed only slightly. Villages still occupy hilltops and tribal women haul household water from the streams below. The men fish the streams with nets and sometimes with hooks, but seldom use elaborate bamboo traps like fishermen living further down the valleys, perhaps because the water is too shallow and the fish too small. This characteristic of distancing themselves from the river derives from practical and, in one instance, spiritual considerations. In the narrow gorges, there is little land to till. Moreover, the steep hills limit the amount of sunlight the valley bottoms receive. For the Lisu tribes, the streams are haunted by malevolent spirits. To avoid confronting these spirits, the Lisu build their homes on the slope of the adjacent valley, and construct a bamboo piping system to convey water around the ridge and into their village.[2]

Each of the half-dozen tribes has its own house design and settlement pattern but, in general, they raise the bases of their houses on stout, waist-high posts, and pigs, chickens, and other small farmyard animals shelter beneath the floors. Each house is an individual unit, walled off from its neighbour by head-high bamboo fencing. If the village is situated on a ridge, houses run on either side of a long main road. A hilltop village is clustered around spirit flags at the summit with houses ranged along a main street and the numerous lanes leading from it.

In the past two decades, many tribes have moved down from the mountains to share the more fertile valleys with Thai Yai (Shan) and Thai peoples. They continue to live at the foot of the valley walls, planting rainfed maize and rice on the slopes leading down to a river's edge. They seldom terrace the land except to plant small vegetable plots, although most vegetables are grown in kitchen gardens in each family's yard. Household and drinking water is channelled along a ditch from a point high on a stream; in some instances, the water can run for several kilometres at a barely perceptible gradient before tumbling into the village pond.

Further down the tributaries, in the regions north of the Chiang Mai, Lampang, Phrae, and Nan Valleys, Thai Yai houses are raised to head height on tree-trunk pillars. They are bunched together, often as far as 1 kilometre from a river, and crops are rainfed. In the four valleys themselves, villages are situated according to the height of the river-banks. In the flatter valleys, they are separated from the rivers by broad expanses of rice fields. Where the river-banks are high, the villages protrude over the rivers and houses are ranged along the banks like beads strung on a liquid string.

The Central Plains

It is in the Central Plains that one finds the most intimate union between Thais and their waterways. A house forms an interface between life on the river and that on solid land. It is regarded as an island in an aquatic environment that periodically threatens homes with an over-abundance, rather than a dearth, of water. House orientation, with the entrance door facing the river, reflects not only the owner's awareness of the river's important role in his life but his abiding concern with its whims and wiles. This preoccupation is at the core of rituals performed during house construction. Prior to erecting house pillars, propitiation rites are conducted to placate the spirit of the land—a *nak* (*naga*), the mythical serpent associated with water.

Although construction materials have changed over the centuries, in most instances, modern Central Plains village houses would be immediately recognizable to someone from the Ayutthaya period. Into the packed earth are sunk the pillars which raise the structure as high as 4 metres above the ground, depending upon the height of the land above the waterway. The space beneath the house provides a work area and a barn for draught and food animals (Plate 26). When the monsoon-swollen river or canal begins to rise, the animals are moved to a detached, partially roofed enclosure or, in severe floods, to bamboo pens erected temporarily on the shoulders of public roads.

A stairway or pathway links the house with the water's edge, and a plank bridge often leads from there to a floating landing. The waterfront location offers easy access to receive guests and goods, berth boats, hail water taxis, and clean produce and clothes. From the landing, the householder casts a net or drops a hook to catch fish. The house interior generally comprises a single large reception room. At night, bedding is unrolled on the floor, mosquito nets are hung from the rafters, and the room becomes a dormitory-style bedroom. The kitchen is situated on a side or back veranda, with only a roof to protect it from the elements. In most modern

26. Houses virtually overhang the river in Central Thailand with the banks serving as a pen for ducks.

village homes, a bathroom is built on the opposite end of the back veranda or in a far corner of a house yard. The bathroom is a concrete block structure resting atop a sealed, concrete-lined septic tank.

The preponderance of Central Plains villages are laid out in the 'ribbon' pattern, with houses hugging both banks of a waterway and the fields stretching away from the homes' back doors. While some Plains villages extend up to 2 kilometres along the river, most reach an optimum length after 500 metres; thereafter, growth takes place laterally, away from the banks. The path running along the land side of the houses becomes the village's main street and a file of houses rises on its opposite side. The village 'aristocracy' inhabits the river-banks by virtue of its longevity and wealth. Those who live further back must perform their household tasks at public or *wat* landings at the periodic openings between river-bank houses.

Although in the past large tracts of trees were cleared for farmland, those along the river-bank were preserved to provide shade and to protect the river-bank against erosion. Thus, from a distance the village appeared as a long, dark-green island in a sea of chartreuse rice plants. Even today it is possible to ascertain the location of a distant village, and the river flowing through it, by looking for a dark-green band of trees.

When a village became prosperous, a Buddhist *wat* was built to face a river-bank or the rising sun (Plate 27) in the belief that Buddha was seated on a river-bank at dawn when he attained enlightenment. As the village became a town, it extended into the river itself with floating houses, generally occupied by fishermen or poorer villagers, rising and falling with the fluctuating river-levels. While in recent years the focus has shifted from the waterways on one side of the

27. Buddhist *wat* typically face the rising sun or the river as here on the Yom.

house to the road on the other, the design of the house itself has remained virtually unchanged.

Some lower Plains villagers have adapted to floods by placing their houses on bamboo floats. Khlong Chao Chet, just west of Ayutthaya, was one such village until the 1980s when the soaring price of bamboo forced owners to abandon the waterways and raise their homes on permanent piles on the river-bank. Another town, Bang Li, situated on low ground in Song Phi Nong district of Suphan Buri province, was built on two levels. During the dry season, its shops were at ground level. When the annual floods inundated the bottom storey, the townspeople moved their goods to the upper floor. Walkways linked the verandas, creating a dry passageway above the flood waters. In the early 1980s, drainage programmes reduced flooding and residents now occupy both storeys year-round (Colour Plates 9–11). In the delta around Bangkok, houses are built on knolls and may be clustered or may sit independent of their neighbours, separated by oceans of rice fields.

Rice Cultivation

The Thai have lived relatively like their neighbours on the mainland of Southeast Asia in an under populated but fertile land, where their requirements for subsistence in the old days were simple and easily obtained. Rice and fish were the staple food and in fact in the vocabulary of the people, the word 'food' is 'ricefish' (*khao pla ahan*) which reflects their main physical needs apart from their cloth which they weave for themselves.[3]

In this manner did scholar Phya Anuman Rajadhon characterize the Thais' dependence upon water. French missionary Bishop Pallegoix underscored Phya Anuman's assertion of the ease with which Thais obtained their food when he commented in 1854 that 'I don't know whether there's any country in the world so rich with its natural resources like Siam. River mud is full of fertilizer. There's no need to nourish the soil, delicious rice can easily grow.'[4]

While the Thais are commonly perceived to grow rice in deeply flooded fields, transplanted rice is a relatively recent innovation. A millennium ago, in portions of the lower valley that were under fresh water for most of the year, wild 'floating rice' grew in abundance. Still found in many areas of the delta, it is a remarkable species whose seed-bearing heads rise above the water on 1.5- to 3.0-metre-long stalks. It is harvested while the water is still high, or when it recedes and exposes the marshy soil.

Until the eleventh century, farmers north of Chainat cultivated dry or upland glutinous rice. Seed was broadcast into a prepared field early in the season and the rains watered it. In succeeding centuries, farmers began building bunds or short walls around their fields to keep the rain-water from draining away. Eventually, this technique evolved into the *muang fai* irrigation systems described in Chapter 2.

As already noted in Chapter 2, the wetland rice common in the Central Plains today may have been growing wild in river basins extending from the Ganges to the Mekong. Mon–Khmer farmers resident in the lower Chao Phya Valley were probably responsible for its propagation, but until the eleventh century, it did not move north beyond Chainat. Between the eleventh and fifteenth centuries, however, wetland rice began pushing dryland rice plantation areas northwards. After the fifteenth century, cultivation of upland glutinous rice was restricted to the far northern hills and to the dry North-east. Wetland rice prevailed in all the valleys where *muang fai* irrigation had been established.[5]

Farmer preference for wetland over upland rice derives from several factors. While yields are somewhat higher than those for upland rice, the principal value of wetland rice is its stability. Once upland rice has been planted in May, the farmer is at the mercy of the heavens until the heavy rains of September. When the rains slacken, as they normally do during July, the crop can wilt and die in the burning sun. By contrast, wetland rice is planted in seed-beds in May before the rains begin falling. After the arrival of the first showers in June, the farmer begins preparing the main field, turning the sun-baked field into a slurry and tamping down the clayish soil to form a hardpan under the field to prevent the water from soaking into the earth below. Centuries ago, this was accomplished by driving water-buffaloes around the fields, their hooves packing the soil into a hardpan and, at the same time, making a thick, nutrient-rich soup of the remainder. By the Ayutthaya period, the plough was accomplishing the same objectives. Only when the farmer was certain that the rainy season had begun in earnest would he transplant the rice seedlings into the main field. By tapping the river and trapping rain, the farmer could ensure the crop had sufficient water through the July dry spell. By controlling the level of water in his field, the farmer could plant two crops per year instead of the single crop permitted by the short rainy season.

Long-grained, fluffy wetland rice also prevailed in the Central Plains because of its greater nutritional value and a perceived aesthetic difference between it and glutinous rice, a subjective judgement by lowlanders who claim that glutinous upland rice is blander, heavier, and a soporific. These considerations seem to have outweighed the greater labour demands involved in planting wetland rice. The more time-consuming transplanting method also meant that a family was limited in the amount of land it could cultivate, a constraint compensated for by higher yields and the opportunity to double crop.

Patterns of plantation changed again during the Ayutthaya period. With the rise of populations in the lower Chao Phya Valley, rice cultivation moved deeper into the southern delta. The development of Ayutthaya as a metropolis and the burgeoning rice-farming population mutually spurred each other's growth. The process accelerated with the shift of the capital to Thonburi in 1767 and then to Bangkok in 1782. Previously forested areas were turned under the cultivator's plough, and the swampy areas were bunded for the propagation of wetland rice. Even then, the low population density meant that vast areas remained unclaimed until the late 1800s when King Chulalongkorn decided to develop rice as Thailand's main export product and initiated ambitious irrigation schemes to achieve his goal. Wetland farming techniques and the transplanted rice varieties remained constant until the 1950s when a revolution in farming occurred, for reasons that will be explored in Chapter 6.

Today, the farmer plants two or, if irrigation water is available, three crops. Between rice seasons, he plants field crops if the soil conditions and irrigation system permit it. In the lower Nan, Chao Phya, and other broad valleys, he cultivates peas and other crops on the long, sloping river-banks which are enriched each year by eroded topsoil deposited by the receding flood waters. Farming has also become highly mechanized. Water-buffaloes, the principal farm draught animals until the 1980s, are now kept as milk or meat animals. The farmer uses an 'iron buffalo' (rototiller) or, if he can afford it, a tractor to till the soil. Motorized harvesters and threshers are specially designed for small, marshy fields.

The farm family also pursues a form of hydroponics, planting vegetables which float on the water or extend long roots into the river-bed. Besides the *bua* (lotus) whose stem and seeds are considered delicacies, farmers plant *phak bung* (swamp morning glory), *phak chi* (coriander), *phak krachet* (water mimosa), *taptao* (*Mimulus orbicularis*), *santawa* (*Ottelia alismoides*), *kachap* (water chestnut), *phaeng phuai* (creeping water primrose), *phak pet nam* (alligator weed), *sanona*, *phak waen* (water clover), and a half-dozen others which have no European equivalents. In addition, they grow *ton kok* (rushes), which are dried and woven into mats.

In the lower delta, these and many other vegetables are sold in the floating markets which have sprung up in the canals (Colour Plate 12). Some of these markets are famous tourist attractions but there are dozens of others known only to villagers. Village women rise early in the morning to load their sampans with produce and by dawn are already paddling to the market. Transactions are carried on in the middle of the canal, with boats clustered together and items passed from boat to boat.

Fishing

In commenting on Thailand's ideal growing conditions for rice, Pallegoix could also have been describing the fecundity of the Chao Phya system in producing an abundance of fish and shellfish. *Pla taphian, pla daeng, pla kaben* (freshwater stingray), *pla ma* (horse fish), shrimp, clams, eels, and other varieties are fried, grilled, dried, salted, or left to ferment for a fish paste or fish sauce condiment, providing the main source of protein and salt in millions of farm families' daily meals.

Thai villagers have developed a wealth of fishing techniques, each specific to a particular variety. These methods demonstrate a patient and observant eye to the species' movements, moods, and personality quirks. Equally evident is indigenous ingenuity in devising nets, traps, and other mechanisms from simple, locally available materials in order to outwit the fish. In 1897, Sommerville encountered one method which is still used in the canals from Ayutthaya to the sea (Plate 28):

In the water, near the bank, a frame is constructed, on which is rigged a large wheel; and ropes and pulleys are there, by which a broad suspended net can be lowered and sunk to the bottom of the river. The fishermen then, in several boats, make a long circuit in the broad, deep river, shouting, beating the water with long bamboos, and ringing gongs so that the frightened fish are driven over and into the net, which their comrades at the rack draw up slowly out of the water. The boats then gather quickly around and load up the fish. We have seen large hauls made by means of this stratagem.[6]

This use of the *yok yaw* net is but one of more than four dozen methods the author has identified while paddling down the Chao Phya and its tributaries. The simplest, a line and hook attached to a bamboo pole, has many variants. In one, a single filament is stretched across a river with a dozen smaller hooked lines suspended from it. In another, a long cord is buoyed by floats; from it dangle a dozen baited-hook lines spaced at intervals. At dusk, the cord is stretched along the edge of shore vegetation and the hooked fish are collected the next morning.

In slow-moving, waist-deep water, farmers drive stakes into the river-bottom to form a circle 5–8 metres in diameter and place

28. The *yok yaw* net hangs from a set of cross-bars levered on a frame which allows the net to be lowered into a canal.

THE CHAO PHYA

tree branches within it to create a small thicket. Fish, lured by the seeming security of the shadowed area beneath the branches, take shelter and are caught either by hooked lines or are driven from their refuge and into nets. More elaborate are the fish fences built across fast-running northern rivers. At the end of the rainy season, a fence of thumb-thick bamboo stakes is driven into the river-bed from one bank to the other; finger-wide spaces permit the river to flow through unimpeded. Two-metre-wide openings are left at a half-dozen intervals in this picket fence, and above them, thatched huts are erected on stout posts (Plate 29). Upstream from the hut, a 'V' of logs is built with the apex pointing downstream. Fish feel their way along the fence until they encounter the opening and the nets or hooked lines dangled by fishermen seated in the huts. This type of fishing is considered a village social event and, much like duck hunters in blinds, men spend lazy afternoons drinking and conversing as they wait for the fish to bite.

Thais also catch fish with a variety of painstakingly crafted bamboo traps (Colour Plate 13; Plates 30–31). The traps normally rest half or completely submerged in the shallows, with a series of complicated openings which allow a fish entry but block its exit. The largest types have trapdoors weighted with large stones. Other traps are actively employed: the fisherman walks through the water, stalking the fish. In one instance, he places a conical basket over a suspected fish lair and extracts the catch through the trap's upper opening. Others, shaped like coal shuttles, are scooped through the water to catch freshwater shrimp (Plate 32).

Nets woven in various sizes and shapes are employed in several ways. Some, like the *yok yaw* Sommerville described, are suspended from frames. Long, small-mesh nets may be stretched across a river or may be grasped at either end by two fishermen who walk through shallow water, sweeping up any fish swimming there. In the swift

29. Shortly after the monsoon rains abate, men begin driving the stakes to build fish fences.

30. One of many intricate bamboo fish traps which are set along the banks of the Wang River.

31. Some of Thailand's finest craftsmanship is applied to the creation of fish traps, like this one on the Yom River.

32. This trap is employed like a scoop with the fisherman walking through a shallow canal in search of his quarry.

upper rivers, men or women dip wide nets affixed to Y-shaped bamboo poles into the water along the river margins (Colour Plate 14).

In the Sirikit Dam reservoir, fishermen work at night from large floating platforms. Twin pylons 6 metres tall rise from the forecorners of a stationary raft; a sturdy frame is hinged to the raft front. Thick ropes run from the frame, over the pinnacles of the pylons and down to winches anchored at the rear of the raft. The winches lower the net into the water and a gas lantern affixed above the net attracts schools of fish. At intervals through the night, the fisherman raises the net and removes the trapped fish. At smaller dams, flat trays made of mosquito netting are suspended from the faces of the spillways. Migratory fish attempting to leap over the dam fall instead into the trays. In the Bhumibol Dam reservoir, the bows of powerful motor boats are fitted with frames with wide nets suspended from them. The boat ploughs through the water, scooping up fish.

One of the more unusual methods utilizes a narrow, 6-metre-long boat called a *rua kradan phi lok* (ghost boat). Attached to the high, right-hand gunwale is a wide white-painted wooden plank whose lower edge trails in the water. Stretching out from the left side of the boat is a broad net. At night, a fish is attracted to the boat by the white board glowing in the dark. It leaps, thinking it can clear the board, and lands, instead, in the boat or in the net. Despite the seeming flimsiness of the premise, it works, a testament to observation and ingenuity.

On all rivers and canals, fishermen cast circular nets with lead-weighted rims which drop into the water in wide circles, trapping the fish beneath them. On the upper rivers, bands of boats prowl the waters. Selecting a site, they draw their craft into a wide circle with their bows pointed towards the centre. At a signal, the boatmen paddle rapidly towards the circle's centre; someone in each boat slaps the water with a stick or a paddle to create a commotion; and, at a shout from the leader, all the fishermen standing on the boat bows hurl their nets at the same time, trapping large numbers of fish (Plate 33).

Although fishing is an informal pursuit, there are several traditional rules and taboos governing it, some of them quite formalized. In his 1837 journal, Edmund Roberts noted that a pair of white wooden pillars were driven on either side of the entrance to Khlong Bangkok Yai and on the Chao Phya above and below Bangkok to indicate that fishing was prohibited within the capital. Similar fish and animal sanctuaries in several provincial capitals discouraged the taking of animal life.[7] While there were no religious strictures against fishing in rural rivers or canals, the practice was considered to be anti-Buddhist and uncivilized and was therefore denigrated by urbanites who considered themselves above such primitive activities. Terwiel notes that this sense of moral superiority is reflected in the writings of urban poets who protest with dismay

33. Gang net fishing on the Ping.

the vileness of the villagers in catching fish, although, as he comments, 'Bangkok people readily bought fish and fish products. The fisherman's *karma* had already been burdened by his deed of killing. In Thai opinion there was no evil attached to the buying and consumption of the catch.'[8] Today, such prohibitions are no longer observed and city fishermen cast their lines from boats or quays at any time of day. For many, it is merely sport fishing, their reluctance to eat the fish from the putrid waters stemming more from health concerns than moral repugnance.

In the past, fishing was prohibited near a *wat* for two days each month. In recent years, many up-country Buddhist abbots have prohibited fishing in canals or weir ponds adjacent to their *wat*. This *khet aphaiya than* (mercy boundary) has been established less for Buddhistic purposes than to preserve the fish from over-exploitation.

Boats

Foreign journals are filled with accounts of the easy familiarity of Thais with waterways: 'A knowledge of the art of swimming is highly necessary to those who inhabit floating houses, and it forms one of the earliest branches of the education of the children. Even infants that are hardly able to walk, may be sometimes seen paddling about in the river, almost as if it were their natural element.'[9]

European writers expressed wonder at the ubiquity of boats of all types and sizes. Sir John Bowring echoed many writers before him with his observation in 1855 that 'Boats are the universal means of conveyance and communication. Except about the palace of the Kings, horses or carriages are rarely seen.'[10] Boats conveyed commerce and people the lengths of labyrinthian rivers and canals.

Every family had at least one boat and various authors estimate that there were more than 200,000 craft in seventeenth-century Ayutthaya alone. The wide range of boat types suggests consummate skill in constructing vessels to carry a wide variety of cargoes, engage in warfare, and operate in waters ranging from turbulent to dead calm.

Nearly all visitors comment extensively on Siamese boat-handling skills. George Windsor Earl marvels at their ability to propel the tiniest of craft without a twinge of fear:

Upon landing, I met Hattee, Mr. Hunter's servant, coming down from the house with something under his arm which looked like a child's coffin; on his approach, however, I discovered that it was a light canoe, about five feet long, and a foot and a half broad, in which he was about to take an excursion on the river.... I felt great curiosity to see how he could possibly get into such a skimming dish.... Having launched the canoe in the water, he steadied himself by holding fast to a post conveniently driven into the bank, and placing one foot gently into the center of his diminutive vessel he gradually drew in the other. I certainly expected to see the boat slip from under him; but he seated himself comfortably, and giving me a knowing look, flourished his double-bladed paddle, and struck off into the middle of the river, the sides of the canoe not being more than two inches out of the water.[11]

Royal Barges

The most magnificent of Thai water craft are the royal barges. For the English speaker, the term 'barge' commonly refers to mammoth river vessels filled with bulk goods and pulled or pushed by tugboats. In Asian usage, the term includes the large, paddled craft that convey royalty to and from ceremonies of state or religion, much in the manner of Cleopatra's barges on the Nile River. These long, narrow, and often sumptuously decorated barges evolved during the early Ayutthaya period as warships, many fitted with cannons whose snouts protruded through the bows. In peacetime, they were the key participants in elaborate royal ceremonial processions (see Chapter 5).

Given their length and narrow beam, the barges would seem ill-suited for battle. They tend to track in a straight line and thus are difficult to manœuvre quickly in a narrow, winding river such as the one embracing Ayutthaya. Although the extant records do not indicate so, it is probable that the larger barges were command centres. Like the great land battles in which princes directed their troops from elephant back and the burden of combat fell on common foot soldiers, the principal barges may have been attended by squadrons of small, easily manœuvrable craft carrying the soldiers who actually engaged the enemy (Plate 34). These small boats would have been highly effective against a land-based invader like the Burmese, slogging through marshy countryside and forced to cross numerous canals and streams.

Even at this early date, the principal barges—which were likely

34. A mural depicts the manner in which small boats were employed in war against a much larger enemy junk belonging to the Cambodian noble, Phya Chin Chantu, during the reign of King Naresuan. (Courtesy National Archives)

reserved for ceremonial purposes—appear to have been very large. La Loubère, writing in 1693, notes that a typical barge was 'composed only of one single Tree, sometimes from sixteen to twenty Fathom [30–40 metres] in length. Two men sitting cross-leg'd by the side one of another, on a Plank laid across, are sufficient to take up the whole breadth thereof.'[12] During the reign of King Narai, a ceremonial barge procession could comprise up to 113 barges and 10,000 oarsmen. The sheer number of paddlers in a population of perhaps 500,000 inhabitants suggests both the fleet's importance in war and the lavishness of the procession in peacetime.

With the Siamese royal fleet destroyed, one of Rama I's initial acts on ascending the throne in 1782 was to order the construction of a new principal barge. Measuring 36.15 metres long, 2.90 metres wide, and apparently modelled on an Ayutthayan original, it was christened *Suphannahong*. In Rama II's reign, a second, larger barge was built and in Rama III's reign, the regal *Ekkachai*, one of the fleet's largest barges, was built. In 1831, Rama III travelled in the *Suphannahong* (Plate 35) to several small village temples to listen to the monks pray.

Although Thailand was no longer harassed by its neighbours, Rama III perceived a threat and initiated a major barge construction programme. Among the many additions were twenty lightweight boats named for lines of poetry. In 1827, Malloch wrote in the Burney Report:

The King has within the last eight or ten months erected an extensive shed covered with tiles about a mile up the Bazar river, nearly opposite the

35. Rama III's Royal Barge. (Private collection)

Prah Klang's house, in which I counted 136 war boats 60 feet long 7 feet broad in the middle and 3.5 feet in the stern, and capable of carrying about 30 men. A similar shed has been erected a very little distance in shore at Bangkok Noi river [canal], nearly opposite the Palace, about a quarter of a mile up, 100 boats of the above description. Immediately above the Palace and on the same side with it, the Government have constructed smaller but similar sheds to the above along the banks of the river.[13]

The Bangkok Noi sheds still house several dozen barges including the flagships of the fleet.

Weather ultimately erodes all boats, and the *Suphannahong* was not immune to its ravages. Rama V had ridden Rama I's *Suphannahong* to present new robes to monks at Bangkok's Wat Arun during Thot Kathin, a ceremony held each October to mark the end of the three-month Buddhist lenten period. Within a few years, however, the barge had deteriorated and Rama V replaced it with a new one which was christened the *Sri Suphannahong* (Colour Plate 15). Carved from a single teak log, the boat, 44.90 metres long, 3.15 metres wide, and 90 centimetres high, was ridden by King Vajiravudh (Rama VI) (r. 1910–25) to his coronation ceremony on 4 December 1911. Maintenance for the forty-five vessels became so expensive that 1929 saw the last procession of the fleet for the next thirty years. In 1961, the Royal Barge Procession to convey the king downriver to Wat Arun during Thot Kathin was revived as an annual event.

The present fleet comprises fifty-one vessels and is divided into five classes descending in size and lavishness of decoration. The first class, which conveys the king and his immediate family or important Buddha images, includes the three principal barges: the *Sri Suphannahong*, the *Sri Anandakaviloka*, and the *Sri Nammachalai*. It is followed by the second-class barges for high-ranking members of the royal family, and the third class for ordinary officials. Fourth-class, undecorated, and unpainted barges transfer the king and his retinue from the ceremonial barges to shore or carry them through difficult or shallow waterways. Fifth-class barges include gunboats with animal figure-heads: *krut* (garuda), monkey, lion, half-lion, half-elephant, horse, and mountain goat. These mythical beasts were the symbols of the military units in the royal processions, and each boat carried a bow cannon.

By 1967, the barges had decayed to such an extent that major restoration work was undertaken. With a budget of B2.5 million, new barges were constructed of *takhian thong* (*hopea odorata*, a type of dipterocarp often called 'ironwood'), a wood more flexible and durable than teak. The original prows were retained and teak was used for the carved decorative sections on the barges' hulls. Further renovation was carried out in preparation for the 1982 and 1987 processions, but presently, the barges lie unused in the Bangkok Noi sheds that serve as a museum.

The boats ridden by lesser royalty were manned by a retinue of paddlers, much like sedan chairs bearing land-based nobles were carried by coolies. The larger boats employed sails and/or a dozen paddlers who sped their owners to riverside *wat* to hear sermons or to visit friends or relatives down a distant *khlong*. A noble could decorate his boat, a right denied even to ministers of state.[14]

For these and the smaller river boats described further on, a special vocabulary evolved to describe modes of propulsion. Thus, *chaeo* meant 'to stand and row', while *phai* meant 'to paddle', and *thaw*, 'to pole'. Among the colourful terms for boat components was the *hang sua* (tiger's tail), for 'tiller'. Boats and their occupants were protected by the guardian spirits which had inhabited the original wood. While, aside from barges, Thai boatmen did not paint eyes on the bows as did Chinese sailors so the boat could 'see' its way through the water, they did, and still do, attach incense, flowers, and ribbons to the bowsprit. By these offerings, they propitiate the female *mae ya nang* spirit which formerly occupied the *takhian thong* tree, and which, by extension, now inhabits and protects the boat itself.

Many traditional boats disappeared around the 1920s as motorized, streamlined craft became popular. Among the more unusual craft was the *rua hang malaeng pong* (scorpion tail). With a stern curved sharply upwards and forked like a scorpion's tail, the 15- to 16-metre boat was poled by four men. Carrying a cabin amidships and another at the stern, it served as a freighter and passenger ship on long journeys along the Ping River. Despite its size, it rode high in the water, gliding easily over the shallow waters of the Tak sandbanks and of the Kaeng Soi rapids, which was upriver from the present Bhumibol Dam.

Two variants, the *rua mae paw prathun* and the *rua mae pa keng*, had a *keng* (flat-roofed cabin) in the middle and another in the stern to shelter passengers. Each was rowed by an oarsman sitting in the bow, facing the stern, an unusual position for a Thai boatman, who preferred to see where he was going. In shallow rivers, the craft were poled. The 10-metre-long *rua sam kao* (three

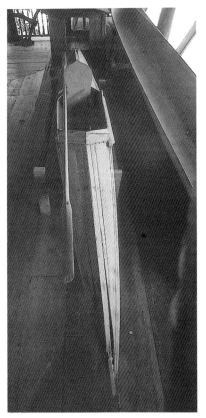

36. A *rua khem* (needle boat) favoured by teenaged boys.

37. A bamboo raft of the type once used in the upper reaches of the tributaries.

steps), with a permanent wooden cabin located amidships, was first used during Rama V's reign to convey noble families up and down the river. Steered by a tiller, it was rowed with oars attached to posts, three on the front and three on the stern deck.

The long, very narrow, late-nineteenth-century *rua khem* (needle boat) (or *rua u*) was a sleek, 3- to 5-metre boat normally built of teak planks and used only for short journeys (Plate 36). It was a favourite among teenaged boys, who raced it on the rivers and canals to impress girls. Kayak-shaped with pointed prows, it was propelled by a single occupant wielding a two-bladed paddle. He sat midway between the prows, his lower body encased by the deck.

Village Boats

The only boats that travelled the northernmost portions of the tributaries were rafts used for single journeys and then discarded since it was virtually impossible to paddle them back upstream (Plate 37). Made of bundles of bamboo poles lashed together with vines—a type of craft still in use today in the north—the rafts offered the advantage of easy construction, space to carry large cargoes, and convenience since they could be dismantled, portaged around obstacles like fallen trees, then reassembled for the onward journey. At the final destination, the bamboo would be sold or used as construction material.

In the lower rivers, permanent boats were built for fishing or to ferry cargo and passengers. In 1968, Phya Anuman identified twenty-nine different paddle boats utilized in the Central Plains.[15] The present author has counted thirty-two types (not including variations on basic designs) on the middle and lower rivers alone. Advanced technology and more durable materials, the incorporation of engines, and new transport demands have raised that number to more than forty. While dugouts have been used since the thirteenth century, most of the more elaborate boats described

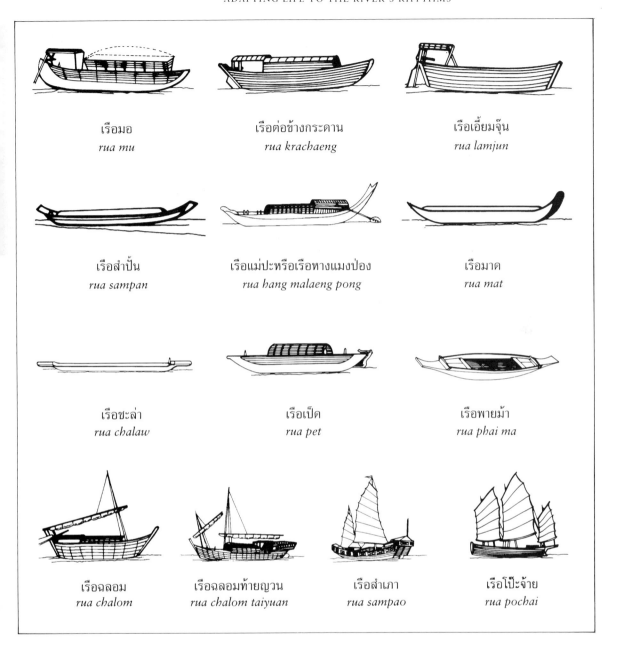

Figure 1. The shapes of various principal boats used in the Chao Phya.

below made their appearance by the mid-Ayutthayan period. In only a few instances in the lower river are boats painted or varnished.

Phya Anuman divided boats into three categories according to their method of construction (Figure 1): *rua khut* (dugouts carved from single logs), *rua mat serm krap* (dugouts with gunwales), and *rua taw* (boats assembled from planks).[16] Today, although substantially diminished in numbers, examples of most types can still be found. Limited space precludes enumerating more than a handful of the most prominent or interesting types.

The simplest *rua khut*, the *rua mat*, is a hardwood log, generally a *takhian thong*, which has been hollowed out with an adze. Its size can be enhanced by heating the carved log over a fire or by placing smouldering rice husks in its interior. Once the wood has softened, the two halves can be prised apart to broaden the boat's beam, a process called *put rua* (open the boat). This type of boat has enjoyed widespread popularity since ancient times because it handles well in many types of water, especially the shallow northern streams. Its thick hull also endures the heavy pounding on rocks that would batter a more fragile boat. Similar in concept, the *rua i-phong* is carved from the trunk of a palm (*ton tan*) tree and used as a fishing or ferry boat in the tributaries.

In the Central Plains, a variety of *rua mat* (Plate 38), 8–10 metres long, is steered with a tiller; it is rowed by four standing oarsmen or driven by an engine. Distinguished by its duckbill-shaped bow and stern, it is among the oldest of Thai boats. It normally transports rice, firewood, and stones but is also used for fishing, village *thot kathin* (presenting robes to the monks) ceremonies, and longboat races. The *rua chalaw* differs from the *rua mat* in having a squared-off bow and stern and by being nearly the same width along its entire length. Made only of teak and used primarily to transport rice, it first appeared in Kamphaeng Phet in the reign of King Chulalongkorn.

Many of the *rua mat serm krap* have disappeared. Carved from a small *takhian thong*, *makha*, or teak log, the *rua mu* (pig boat) bulges amidships, but narrows gracefully to small, pointed prows. One of the oldest native boats of Thailand, this paddled or poled boat is popular for short-distance travel or fishing because it is stable and comfortable. One variant, the *rua plu*, is somewhat larger; the other, the *rua kulae* (or *rua pulae*) is smaller. The *rua kulae nok pak* is shaped like a large *rua kulae* but a *krae* (running board) is added to the sides to provide a narrow platform along which the poleman can walk.

38. A *rua mat* with an outboard engine on the middle Yom.

Dating from the Ayutthaya period, the *rua phai ma* is a dugout of *takhian thong* with a double bow (*hua kratai*; rabbit head). Six to seven metres long, it transports rice and grass from the fields and is sailed or paddled from a seated or standing position. Passengers are protected by a *khaiyap* (curved roof). The large *rua kanya* takes its name from its *kanya*, a peaked roof made of woven *attap* or screwpine leaves which shields passengers. Its distinctive feature is a central mast secured by ropes attached to the bow and stern. When the roof is covered with red cloth, it is called a *rua si sakalad* and is used as a royal barge. The *rua nua*, similar to the *rua kanya*, formerly carried cargoes or the cattle of Indian herders in the northern provinces.

The following boats were once seen in great numbers on the rivers. The very long *rua waet* had a large, high stern, and was protected by a *keng*. It was said that Somdet Chao Phya Maha Si Suriyawong, the regent during the reign of Rama V, modified it from a Vietnamese boat, the *rua wae*. The *rua sampan keng pung* was also inspired by a Vietnamese boat, the *rua chaew*. Between 28 and 30 metres long, and paddled by standing oarsmen, it had a *keng* amidships. Designed to carry royalty and high-ranking officials between Samut Prakan and Bangkok, it was likely used to convey foreign envoys like Sir John Bowring and others from their ships moored at the river-mouth to the royal courts in Bangkok. It disappeared after the reign of Rama V. The *rua mu*, a large dugout with gunwales and a *khaiyap* at the stern, transported cattle, straw, or rice husks along the upper reaches of the Chao Phya. The six-oar *rua kechai*, somewhat larger than a sampan, transported fresh fish from Paknam to the markets in Bangkok.

Most boats in the third category, *rua taw* (boats assembled from planks), were introduced to Thai waters in the last century. The use of planks allowed craftsmen to construct larger vessels with a wider variety of shapes and uses. Perhaps the oldest type, the *rua taw khang kradan* can be 15 metres long and is roofed fore and aft. Traditionally used as a floating store, many are now floating homes. On both sides of the boat, above the level of the gunwales, walls of two or three planks, each one *sok* (cubit) wide, are erected. A curved galvanized iron roof covers the boat. The *rua pochai* (also called *rua po*), a lighter version of the *rua taw khang kadan*, is a cargo transporter. A mast with sail is stepped amidships to propel it like a *rua sampao* (junk).

The Chinese-influenced *rua sampan* first appeared in Bangkok in the late 1700s. A lower-river teak boat, its name 'three planks' describes the construction of its hull. Sometime in the nineteenth century, however, the three planks were increased to five to broaden and lengthen the boat. The largest type is 16 metres long, normally has a *khaiyap* roof in the middle, and carries passengers and/or cargo; at least two oarsmen are required. A second type, 6–8 metres long, transports passengers. The smallest sampans carry vendors selling food and sundries. Designed for monks, the long and narrow

THE CHAO PHYA

rua sampan phrio (streamlined sampan) enables them to complete their alms rounds quickly.

The *rua pet* (duck boat; so-called because, like a duck, it sits high in the water) resembles a sampan with rounded prows. A northern valley boat invented in the 1930s for fishing and mercantile purposes, the *rua pet*'s flat bottom renders it suitable for use in shallow rivers. A seagoing relative, the *rua pet thale*, has higher prows. The name and design of the 5-metre-long *rua bot* are derived from the common European lifeboat. Native to the central region since the reign of King Chulalongkorn, it is used over short distances and in seas with low waves. A relative, resembling a pointed-bow, square-stern Western-style rowboat, is rowed rather than paddled.

The Chao Phya Workhorses

Although classed as *rua taw*, the elephantine barges are in a special class by themselves. The most common, the *rua taw krachaeng* (or *rua krachaeng*) (Plate 39) was, until the 1970s, the principal workhorse of the lower Chao Phya, Tha Chin, and Pasak cargo fleet. Measuring 10–18 metres long and 5.5 metres wide, and constructed of single planks of second- or third-grade teak (first-grade teak was reserved for export), the gracefully curved hull was similar to the larger *rua iamchun* (described below). The name *krachaeng* describes a roof of *attap* or screwpine leaves woven into mats, although on the poorer models bamboo and other leaves were pressed between

39. *Rua krachaeng* barges waiting to be loaded with cargoes near Bangkok.

bamboo lattices for use as roofs. In recent years, galvanized iron has become the preferred roofing material. When loaded, it rides so low that its fore and aft decks are awash.

The *krachaeng*'s size is limited by several considerations. When studying the canals, one is struck by their extreme shallowness. These canals are not deep grooves in the earth but are mere scratches, meaning that the barges virtually glide over the surface rather than ploughing through deep waters. This factor limits their size. Sand barges could be up to 22.0 metres long but their width was restricted to 5.5 metres, a few centimetres less than the watergates along the canals they often plied. The larger models could carry 900 bags of rice. In the nineteenth century, they were poled or paddled up and down the canals and main river. With the advent of motorized tugboats, they lost their independence, transformed from locomotives into so many anonymous railcars towed in a long line, with the barge owners paying the towboat captain a fee for his services.

The techniques of barge construction are rapidly being lost as new materials are introduced. Boat-yards along the river-banks above Bangkok once required 3–4 months to create a large *krachaeng*. The planks were suspended above fires to curve them before they were fastened to the ribs of the boat's skeleton (called a *kraduk ngu* or 'snakebone') with wooden pegs which, when moistened by the river, would swell to fit snugly and would not rust as did nails. In the final step, called *tok man*, raw wet thread was pounded into the cracks between the planks as caulking (Plate 40).

The large, broad-beamed *rua iamchun* (or *rua chalom*) is the graceful Rubenesque beauty of the lower river. Similar to the *rua krachaeng*, its distinguishing features are its small T-shaped roof and support on the aft deck and the tall, slim twin rudders at the stern (Plate 41). It is towed behind a motor boat, poled, or rowed standing, and is used to haul cargo, usually serving as a lighter

40. The *tok man*, the finishing touch on a recently repaired *rua krachaeng* in which resin-soaked thread is hammered between the thick planks as caulking.

41. From any angle, the *rua iamchun* or 'salt boat' is a graceful vessel. It is now used primarily as a lighter for cargo transfer between ships and shore.

between ships moored in midstream and the shore. The *iamchun* traces its lineage to a large seafaring Chinese vessel with twin rudders and two masts stepped in the foredeck. In the Tae-chiu dialect, its name means 'salt boat' as does its alternate Thai name, *rua klua*, because until recently, it carried salt along the canals from the flats in Samut Sakhon to Bangkok. In the nineteenth century, when the hot-season river-level dropped and the invading sea turned the river salty and undrinkable, the *iamchun* would be sent upriver to be loaded with fresh drinking water for sale in Bangkok.

River Boats in Recent History

Until the completion of the Bhumibol Dam in 1964, boats ran regular routes between Bangkok and Chiang Mai. The most formidable obstacle on the Ping River were the Kaeng Soi rapids; on the Nan, the Kaeng Luang rapids; and on the Yom, Kaeng Sua Ten. The only large boats capable of manoeuvring through them were the *rua hang malaeng pong* or scorpion tail boats. Rowed or poled, the boats could haul 5 tons of passengers and cargo from Bangkok to Chiang Mai (Ping), to Nan (Nan), or to Phrae (Yom) on a journey that could take up to six weeks. After the dams were completed, the rivers were supplanted by roads as the main north–south transport routes.

The mid-nineteenth century witnessed the introduction of motorized boats. The 'steam launch' was a tug and transport boat in its own right, filling the Chao Phya from Nakhon Sawan to the sea with its steam, smoke, and soot. The early 1900s saw its consignment to history as diesel-powered internal combustion engines assumed its duties. The storeboats described in Chapter 3 disappeared in the 1970s, but some of the smaller ones have found

ADAPTING LIFE TO THE RIVER'S RHYTHMS

42. The sampan as a floating noodle shop.

43. A sampan coffee-shop.

44. A *rua hang yao* or long-tailed boat propelled by a motor-car engine mounted on a pivot. It operates as a canal taxi boat.

new life in rural communities. Serving as mobile post offices, medical clinics, libraries, and playrooms, they follow a regular route, calling at riverine villages for several hours at a time. The *rua yon*, a 7-metre-long, roofed motor boat formerly used as a small store-boat, now ferries tourists along the Thonburi canals.

Sampans still travel from door to door, offering sweets or noodles cooked on board on small charcoal stoves (Plates 42–43). Others sell sundries, buy scrap paper or metal, or carry steaming vats for dyeing clothes. In the main river, they carry compressors and helmets for divers who scour the river-bottom for lost items they can sell to junk dealers. Sampans have also been used as floating brothels.

Until the 1980s, it was still possible to see the small boats under the bridge over the Saensap Canal at Bangkok's Pratunam intersection. With their mosquito netting shielding the occupants from insects and the stares of passers-by, they gave new meaning to the terms 'floating world' and 'pleasure craft'.

A purely Thai invention, the *rua hang yao* (long-tailed boat) overcomes the problem of manœuvring in narrow, shallow canals by balancing a motor-car engine atop a vertical water-pipe. From the rear of the engine extends a long drive shaft that ends in a propeller. The arrangement allows the boatman 270 degrees of turning radius, enabling him to virtually pivot on the spot. The boats serve as commuter buses, travelling regular routes along the canals and conveying goods and passengers (Plate 44).

While the long-tailed boat represents an ingenious adaptation of technology to river conditions, the thrust of most twentieth-century development has focused on subverting nature's patterns and on substituting human for natural values. Before examining the manner and consequences of river transformation, it is necessary to explore how changing spiritual values have impacted on the rivers.

1. Ernest Young, *The Kingdom of the Yellow Robe: Being Sketches of the Domestic and Religious Rites and Ceremonies of the Siamese*, London: Archibald Constable & Co., 1898; reprinted Kuala Lumpur: Oxford University Press, 1982, pp. 25–6.
2. Paul Lewis and Elaine Lewis, *Peoples of the Golden Triangle*, London: Thames and Hudson, 1984, p. 254.
3. Phya Anuman Rajadhon, *Essays on Thai Folklore*, Bangkok: Social Science Association Press of Thailand, 1968, p. 31.
4. J.-B. Pallegoix, *Description du royaume Thai ou Siam*, reprinted Bangkok: D. K. Bookhouse, 1976, p. 3.
5. Tadayo Watabe, 'The Development of Rice Cultivation', in Yoneo Ishii (ed.), *Thailand: A Rice-growing Society*, trans. Peter Hawkes and Stephanie Hawkes, Honolulu: University Press of Hawaii, 1978, pp. 8–9 and 12.
6. Maxwell Sommerville, *Siam on the Meinam; From the Gulf to Ayuthia*, London: Sampson Low, Marston and Company, 1897; reprinted Bangkok: White Lotus, 1985, p. 107.
7. Edmund Roberts, *Embassy to the Eastern Courts of Cochin China, Siam and Muscat during the Years 1832–3–4*, New York: Harper and Brothers, 1937, p. 236.
8. B. J. Terwiel, *Through Travellers' Eyes: An Approach to Early Nineteenth Century Thai History*, Bangkok: Editions Duang Kamol, 1989, p. 207.
9. George Windsor Earl, *The Eastern Seas, or Voyages and Adventures in the Indian Archipelago, in 1832, 1833, and 1834*, London: W. H. Allen, 1837; reprinted Singapore: Oxford University Press, 1971, p. 22.
10. Sir John Bowring, *The Kingdom and People of Siam*, London: Oxford University Press, 1969, Vol. 1, p. 402.
11. Earl, *The Eastern Seas*, pp. 160–1.
12. Simon de la Loubère, *A New Historical Relation of the Kingdom of Siam*, London, 1693; reprinted Kuala Lumpur and Singapore: Oxford University Press, 1969 and 1986, p. 41.
13. D. E. Malloch, 'Extracts from My Private Journal Relative to Events Which Happened after the Departure of Captain Burney in 1826, to the Period of My

Quitting Bangkok on the 20th March 1827', *The Burney Papers*, Vol. 2, Pt. 4, n.d., p. 227.

14. Letter by Mgr. Bruguiere, dated Bangkok 1829, published in *Annales de l'Association de la Propagation de la Foi* [Annals of the Association of the Propagation of the Faith], 1831, Vol. 5, p. 163.

15. Sathien Koses (pen name of Phya Anuman Rajadhon), 'Ban Tuk Chu, Le Chanit Rua Tang Khong Thai' [Descriptions of Names and Types of Thai Boats], in *Fuang Khwam Lang* [Recollections], Bangkok: Suksit Siam Press, 1968, Vol. 2, pp. 246–9.

16. Oral sources: Nid Hinshiranan, former Director of the Town and Country Planning Department of the Ministry of Interior; Paitoon Khaomala, Director of the Witthayalai Kan Taw Rua (Ayutthaya Boatbuilding School); and Phuthorn Bhumadhon, boat historian and Director of the Wat Yang na Rangsi Boat Museum, Lopburi.

5 The Spirit World and Cultural Life

Water appears to have existed before creation and, as in all Tai creation stories, water is the most salient natural element. Trees and forests are not mentioned. The words lum (wind or air) and phun (rain) appear but their creation is not mentioned.[1]

OF the four elements which comprise the Thai universe—earth, air, fire, and water—water is pre-eminent. As is apparent even in a cursory study of Thai society, water permeates every aspect of Thai life, shaping perceptions, values, and, ultimately, culture. It is natural to assume that the reverence paid to water would be extended to rivers, of which water is the prime constituent. On the contrary, rivers have never been regarded with the same mystical reverence in Thailand as, for example, in India, from which many Thai beliefs derive. Instead, the principal emotion displayed towards them (and nature in general) is grounded in a basic fear of their power, a perception which has both positive and negative consequences. On the positive side, the taboos that grew out of the fears kept the populace from damaging the rivers. However, without a mystical appreciation of rivers and their power to sustain a population (a belief that a river was an entity to be embraced rather than shunned), the importance of preserving them or treating them as special was not instilled in the Thai ethos. Were the ancient taboos still efficacious, such lack of reverence would not pose a problem. Unfortunately, this is no longer the case. While water still engages the Thai consciousness, the taboos protecting the rivers have lost their potency, leading to the abuses which will be examined in Chapter 7.

To understand this dichotomy and its effects, it is necessary first to examine the symbolism of water, and second to study how rivers have figured in Thai philosophy, spirit worship, and the arts.

Thai Attitudes towards Water

In explaining the genesis of the world, Thais are in remarkable concordance with most of earth's earliest peoples. Cultures as widely dispersed as the Judaeo-Christians (Noah) and animistic and Buddhist societies as remote as the Tibetans hold that the earth either began with a cataclysmic flood that provided the elixir of life, or swept away evil so that the world could begin anew. In Thai creation myths, water is regarded as the mother of life, pre-dating the birth of all other natural elements. As Wilaiwan Khanittanan notes: 'In most Tai languages these stories begin with a great flood under which the whole world is submerged, and after the water recedes from the land, a gourd is found on a mountain. The gourd, which contains the human races, is struck open by a god.'[2] In tribal cosmology, water plays almost exactly the same role. For example, Lisu mythology tells of an ancient flood which left only two survivors, a boy and a girl who took refuge in a large gourd. Their children became the parents of the various tribes.[3]

As in other cultures, this awe in the presence of a superior force ultimately bred a belief in animistic spirits to explain nature and to formulate means of ensuring that they supported rather than harmed human endeavour. The most prominent among these spirits was the *nak*, a serpent deity borrowed from Hindu mythology but found in various forms throughout Asia; the dragons of China are but one example.

The precursor of the *nak* is the *nguak* (*nang nguak* in Central Thailand), a naiad or water-nymph whose cult appeared in northern Thailand prior to the twelfth century. While in northern Burma the *nguak* is believed to be the spirit of a drowned person, to the Lao, northern Thais, and Khmers, it is a serpent deity worshipped for its magical powers. It inhabits freshwater rivers, lakes, and even wells, especially the portions shaded by tree boughs. In Central Thailand, the *nguak* has been transformed into a crocodile king named Chalawan, ruler of the fabled river-bottom city of gold and gems called Muang Badan (see p. 116). Villagers near Nakhon Sawan believe that an enormous crocodile, Ta Khe Chao, is the servant of the Phra Phrom (Lord of the Land).

The *nak* myth originated in India where the serpent-god is known as the *naga*. It traces its etymological roots to the Sanskrit word 'nagna' which means 'naked' and refers to its lack of fur, feathers, or other adornment. Living in a nether world of cisterns and caverns filled with precious gems, the *nak* is associated with rivers. Like them, it begins life underground in the bosom of Mother Earth, the creator of life, and undulates across the earth, its shape describing both the course of a river and the arc of river waves. It is also the 'keeper of the life-energy that is stored in the terrestrial waters',[4] an apt description of the life-force it imparts to the crops it irrigates. In some instances, the *nak* is regarded as the female creative force, or as a seductress whose wiles wear away the strength of solid rocks and earth.[5]

In Hindu mythology, the chief *nak*, Phya Nak, drank all the water of the world. Angered by his impertinence, Vishnu ordered the *deva* (angels) to tie Phya Nak to Phrasumen, the cosmic mountain, and squeeze it until the water was restored to the world and its people. Thus, the water regurgitated by Phya Nak is regarded by Thais as *nam amarit* (holy water), and those who drink it gain eternal life. The *nak* is also associated with the final meditation by the Buddha as he strove to reach enlightenment. When the earth goddess, Mae Torani, wrung out her wet hair, to drown tempters and evildoers (Plate 45), the *nak* coiled itself under the unheeding Buddha to raise him above the floods and then spread the hoods of its seven heads to shelter him from the water cascading from Mae Torani's locks.

The mythical *nak* that arrived in Thailand via the Khmer courts is regarded as a benign dragon symbolizing mineral wealth (gems, gold, crystal) and inhabiting several realms, including the waters, the earth, and the sky. It is revered as a minor deity and is honoured during Songkran (see below), one of Thailand's two principal water-related festivals, when Thais beseech it to send abundant rain during the coming rice season. In its most common representation, it streams down the roof edges of *wat* buildings, a decorative element referred to as *nak lamyong* or *nak sadung* (Colour Plate 16). A truncated form called a *tua ngao* decorates ordinary house eaves as a symbol of happiness, wealth, and Buddhist piety. Thai-Chinese also regard the *nak* as a potent water symbol, and the Thai-Chinese community at Nakhon Sawan honours it in a twelve-day festival in late January or early February.

Water Festivals

Phya Nak and its cohorts are essential participants in Songkran, an ancient Indian rite conducted to ensure abundant rainfall. Songkran reached Thailand via Burma in the eleventh century, during the reign of King Anawrahta (1044–77) of Pagan. Still observed in Burma, it has derivatives in Laos, Cambodia, and southern China. In Thailand, Songkran formerly marked the beginning of the Lunar New Year and was celebrated on the first day of the waxing moon in the fifth month (April). After the 1932 Revolution, the date was fixed at 13 April. The term 'Songkran' describes the sun's passage from one astrological house to another, and is seen as a means to explain and avert drought.

Its celebration involves seven heavenly sisters, concubines of the god Phra In (Indra), and daughters of a deity named Kapila-brahma, who descended to earth to ask riddles of a seven-year-old prodigy, Dharmapala Kumar. If the boy failed to answer the riddles within seven days, his head would be cut off. If he succeeded, the god would sever his own head.

Six days passed while the boy struggled without success. Preferring suicide to decapitation, he lay under a tree contemplating how to kill himself. Overhead, a female eagle asked her spouse when they would next eat. 'Soon', he replied, and explained why the wretched child below would soon be dead. The wife asked her husband for the riddles and their answers. Overhearing them, the boy sped to Kapila-brahma, and blurted out the answers. Kapila-brahma agreed to fulfil his vow but his head contained magical

45. Mae Torani wrings out her hair to drown the demons attempting to distract Buddha from his meditations.

The upper Nan River flows through a pristine forest.

A very large *muang fai* weir just below Chiang Mai.

3. A diorama at an Ayutthaya museum—a purported view from Ayutthaya's ramparts—depicts the Pasak River on the left joining the Chao Phya on the right to flow to Bangkok with foreign merchant ships anchored just below the city.

4. A modern *rua chai* with a cannon protruding from its bow.

The Portuguese Santa Cruz Church on the Thonburi bank is one of many monuments marking the enclaves of early settlers along the river.

From Phu Khao Thong (the Golden Mount), Khlong Banglamphu, defending the eastern boundary of Bangkok, intersects Khlong Mahanak running to the right.

7. The former river-mouth island dominated by Wat Phra Chedi Klang Nam is now firmly connected to the Chao Phya's western bank.

8. One of many old derelicts that recall an era when the river was the kingdom's chief highway.

The two-storey houses in Song Phi Nong during the dry season.

10. The same houses during the floods when the residents simply removed their goods to the upper floors and carried on as before.

11. The houses today after diking and drainage programmes rendered the town dry year-round.

12. The best known of the floating markets daily draws myriad boats to Khlong Damnern Saduak, west of Bangkok.

3. The positioning of a sturdy trap for larger Yom River fish suggests familiarity with the species' habits.

4. One of many nets used in fast-running rivers.

15. The swan-like figurehead of the *Sri Suphannahong*.

16. A *nak* slithers along the *wihan* roof in a Buddhist *wat*.

17. Monks receive alms at a landing in a Bangkok canal.

18. A spirit house on the river.

19. A royal warship is depicted in a mural at Wat Phra Kaew.

20. A bugler (*inset*) in the Royal Barge Procession as it passes the Grand Palace.

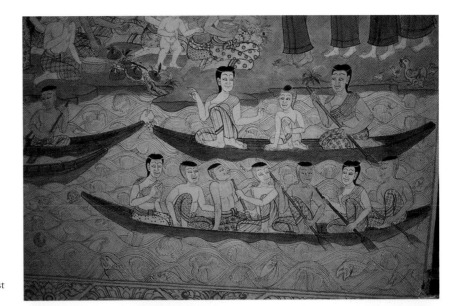

1. A stylized, Ayutthayan depiction of boats and boatmen by an urban artist at a museum in Ayutthaya.

2. Few of the hundreds of water-wheels which formerly lifted water from the Ping to the fields are still in use.

23. The powerhouse of the Bhumibol Dam.

24. Rice harvest made possible by irrigation.

5. Underbrush is burnt during the winter months to plant crops.

6. Giant mimosa covering the Bhumibol Dam reservoir banks extend far into the water.

27. A fleet of dredgers at work in the Chao Phya at Bang Sai.

28. With the loss of taboos, the spirit house and cloths, which formerly protected trees from human harm, have lost their efficacy.

29. Returning the canals to their former roles, boats convey passengers to a variety of destinations.

30. Schoolboy Siriwat Ploydam's imaginative boat transforms rubbish into fish food, one of many drawings in a Magic Eyes campaign to make children more aware of the river's importance.

31. The rototiller or 'iron buffalo' has largely replaced the water-buffalo for tilling fields.

32. The challenge is to increase yields on existing land to halt the destruction of watersheds.

destructive properties and would burn the earth to a cinder if it fell to the ground. Were it tossed into the air, the sky would dry so completely the rain would never fall; if cast into the water, the oceans would evaporate. The gods decided that the head would be placed in Gandha Dhuli cave at the foot of the holy Mount Kailash, thereby averting catastrophe. Each year on 13 April, one of his daughters honours him by carrying his head in a procession while the gods feast.

Up to seven *nak* take part in Songkran. Their exact number determines the amount of rain which will fall during the coming year. Contrary to expectation, the number is in inverse relationship to the amount of rainfall; that is, the more *nak*, the less rain. It is believed that a large number of *nak* would waste time playing whereas one or two *nak* would concentrate on getting the job done. Another interpretation suggests that a large number of *nak* would, like Phya Nak, swallow up all the water, leaving none to shower the earth.

In Bangkok, on the day before Songkran, the kingdom's second most sacred Buddha image, the Phra Buddha Sihing, is paraded around the old royal city. Buddhists lining the route fling bowls of water to wash the residue of the old year from the image. It is then installed in Bangkok's Sanam Luang, the oblong field adjacent to Wat Phra Kaew, where worshippers continue their obeisance throughout the day. The crystal Phra Setang Khamani, which was placed in Chiang Mai's Wat Chiang Mun by King Mengrai after he subjugated Lamphun in 1281, is said to command the clouds. During Songkran, it is carried in procession around the city and water is poured over it to bring rain during the coming rice season. It is also propitiated in drought years and is thought to protect the city from fire. In villages, the principal images are removed from the *wat* and bathed by the monks and the people.

In the North, Songkran eve is spent with friends. At dawn, young men shoot guns and firecrackers to drive away evil spirits associated with the previous year. In the North and the Central Plains, residents clean their houses and kitchen utensils, and wash their clothes and bedsheets to purify themselves for the coming year. Thais then sprinkle water on each other to wash away bad luck and bestow blessings. As the Songkran celebration requires copious quantities of water, it is to the rivers that celebrants go. As the day progresses, the volume of water and the force with which it is propelled increase until the event degenerates into boisterous water wars involving buckets and hoses. Youths drive around the larger towns in pick-up trucks loaded with barrels of water which they generously lavish on passers-by (Plate 46). At the same time, worshippers release fish into the rivers and ponds and free birds from cages, meritorious deeds whose goodness will hopefully redound in the new year. In the courtyards of northern *wat*, devotees build dozens of small *stupa* of sand carried from the river and place small flags at their summits. Their gift raises the *wat* ground, replacing all the soil inadvertently carried out of the

46. Songkran erupts into a full-scale water battle on town streets.

compound in shoes and cuffs during the previous year.

Thailand's second most important festival, Loy Krathong, also has roots in antiquity. In the Lanna kingdom seven centuries ago, the Loy Khro (float away misfortune) ceremony was presided over by the king. He placed a personal object like a piece of clothing in a banana tray and launched it into the Ping, symbolically floating away Chiang Mai's ill fortune and, by extension, cleansing the entire Lanna kingdom. In the afternoon, the king would pour scented water on an important Buddha image associated with rain-calling, such as the Phra Buddha Sihing at Wat Phra Singh and the Phra Setang Khamani at Wat Chiang Mun.

Loy Krathong, Loy Khro's successor, is celebrated on the full moon night of November as millions of tiny *krathong* (boats) are launched on the waterways (Plate 47). A slice of a banana tree-trunk forms the *krathong* hull which is beautifully decorated with flowers and leaves. As a final touch, a candle and incense sticks are installed like masts in the centre (Plate 48). As the moon rises, the celebrant places a few small coins, fingernail clippings, and a few hairs in the *krathong* to symbolize his or her ills. He lights its candles and incense, prays for success in the coming year, and places the tiny boat in a river or lake, creating an enchanting spectacle of light shimmering on the water. As the rite also blesses lovers, many couples launch joint *krathong*, hoping for the success of their union.

Popular tradition contends that Loy Krathong was initiated by a young wife of King Ramkamhaeng to relieve her husband's cares. As he relaxed with friends on a raft on the Yom River, she went upstream and launched a candle-lit *krathong* she had made. The king and his courtiers were so charmed by the *krathong*'s beauty that in succeeding years they began making their own, and the custom spread. Myths aside, it is more likely that the festival

47. In the northern version of Loy Krathong called Yi Peng, celebrants launch *krathong* into the rivers and waterways.

48. *Krathong* with their candles and incense sticks.

originated in India and arrived in northern Thailand via Burma. There are similar festivals in India, China, Cambodia, and Laos but they occur in different months and are celebrated as forms of moon, rather than river, worship. In Thailand, Loy Krathong honours Mae Khongkha, the 'Mother of Waters' (see p. 108), asking that she bring abundance from upstream and carry away misfortunes on her journey to the sea. The rite also asks her forgiveness for harm or pollution during the previous year and implores her to provide her life-sustaining bounty in the coming year.

Bangkok formerly celebrated an annual royal 'Loy Krathong' at the Royal Landing near the Grand Palace. Princes and ministers vied to make the most beautiful *krathong* in the shapes of lotus flowers, junks, and other objects, some large enough to accommodate musicians and comedians who entertained audiences lining the river-bank. Scholar Phya Anuman Rajadhon recalls seeing King Chulalongkorn arriving at 9.00 p.m. to launch his *krathong* which were miniature replicas of the royal barges. In this period, Loy Krathong was celebrated twice each year, with huge *krathong* floated in October, and miniature royal barges launched in November. The king and his court viewed the exhibition while dining on a large floating pavilion anchored in the river; fireworks ended the celebrations well past midnight.[6]

Yi Peng, the northern version of Loy Krathong, is celebrated in November and in the same manner but its origins differ. Legend says that in the fourteenth century, Haripunchai (Lamphun) was struck by a cholera epidemic. To escape its fury, the inhabitants moved to Hong Sawasdi (Burma). When the epidemic abated, most Thais returned but many stayed behind. Missing their loved ones, sorrowing Haripunchai residents placed gifts in *krathong* each full moon night in November, and floated them on a branch of the Ping River in the hopes they would reach their lost relatives.

THE CHAO PHYA

The festival was called *prapheni loy khamot* (floating ghosts) because the *krathong*'s candles created shadows that resembled forest ghosts (*khamot*) whose red eyes glowed in the dark. *Khamot* is thought to be derived from the Khmer word *khmoch* meaning 'corpse'.[7]

Rituals

Thai rituals blend elements of animism, Brahmanism, Hinduism, and Buddhism in a manner that often defies division into distinct categories. None the less, water plays an integral role in nearly every royal state ceremony, Buddhist ritual, and lay rite of passage. Poured like a stream, it can signify a wide variety of emotions ranging from the transitoriness of life to a declaration of fealty to one's superiors, to public avowal of intention to perform an important deed. State ceremonies are conducted on rivers or utilize river water. Water is also a vital element in propitiation rites to bring rain and halt floods.

For a royal coronation, water is gathered from several sources including the major rivers. Once blessed by monks, this lustral water is used to anoint the new king in the Phithi Pli Kam Tak Nam purification rites. For King Bhumibol's coronation in 1950, water was gathered from eighteen sources. In 1987, to celebrate his fifth twelve-year cycle (sixtieth birthday), water was gathered from five sources, including the headwaters of the Ping River.

In the Phithi Lang Nam Thaksinothok rite, water is poured on the ground to signify determination to perform a deed. In such a manner in 1590, King Naresuan declared his resolve to liberate Thailand from Burma (Plate 49). A mural in Wat Suwandararam at Ayutthaya depicts this event as he calls upon the earth goddess, Mae Torani, to witness his act, a sacred vow that cannot be rescinded. Similarly, at his coronation, a king pours lustral water on the earth to signify his determination to rule with righteousness, an act which marks his transformation from a prince to a fully installed monarch.[8]

Until 1932, water was used to proclaim fealty to one's sovereign. For the annual Tu Nam ceremony, waters from rivers in scattered corners of the kingdom were blended at the Chapel Royal, and sanctified by the chief Brahman and Buddhist monks. Sacred swords were dipped into the liquid while ancient Brahmanic scriptures were chanted. Court officials, and in some instances court ladies, were required to drink the libation to pledge absolute loyalty and incorruptibility. Soldiers underwent a similar rite each month to ensure their loyalty. It was believed that if the imbiber hiccuped, sputtered, or choked, it was revealed beyond doubt that he was either disloyal or was contemplating treason. If suspicion was strong enough, the unlucky courtier would be executed. The large number of usurpations of royal power throughout Thai history suggests that the rite may not have carried the force that it

49. King Naresuan pours water on the ground in the Phithi Lang Nam Thaksinothok rite depicted in a mural at Ayutthaya's Wat Suwandararam.

should have. On the other hand, it may have dissuaded some would-be revolutionaries. The ritual was discontinued after the 1932 Revolution that changed Thailand from an absolute to a constitutional monarchy, but since then has been performed on special occasions. The rite was also performed by vanquished foreign kings to signify submission. At the conclusion of a battle, the victorious king would pour water over his sword and demand that the defeated monarch drink it. Should he break his oath, the same sword would wreck retribution upon him.

Sukhothai kings presided over the Brahmanic Phratcha Phithi Phirunsat rites to produce rain in drought years. The rituals were conducted in an enclosed area since they involved a sculpture of a couple copulating in front of a pool holding *deva*, *nak*, and fish. In the past, Buddhist monks and Brahmans chanted for one week while the king was constantly bathed and was not allowed to cohabit with females.[9] A Chiang Mai rain-calling ritual invokes the region's guardian spirits including those of former kings, Lan Na Inthakhin, Pu Sae Ya Sae, Suwanna Khamdaeng, and Mengrai. In mid-May, during the Inthakhin ceremony at Wat Chedi Luang, the Phra Fon Saen Ha rain-calling image is carried through the streets.[10]

Rites of Passage

Thai rites of passage are Brahmanic in origin but over time have become suffused with Buddhist elements. Water is an integral part of the normal rites of passage—birth, coming of age, marriage, and death—celebrated by all Buddhist Thais. For royal children, there were ten ancient rites, of which only three (four if the Buddhist ordination rite is included) are still observed. In the Nammongkon (Phithi Mongkon Kanchu Derm) ceremony, performed on the twelfth day after birth, the child is given a name. For Sikhapanamongkon (Phithi Mongkon Kon Phom Fai Wai Juk), at the end of the child's first month, a lock of the child's hair is snipped to mark his or her acceptance into the world. Chulakantamongkon (Phithi Mongkon Kon Juk Tat Muay Phom), during which a child's topknot is shaved off, marks the child's passage into puberty.

Of those which have disappeared, the Mahathitmongkon (Phraratcha Phithi Long Tha Son Wai Nam) ceremony was reserved solely for the immediate children of monarchs and was one of the most elaborate of royal rituals. The term translates as the 'auspicious rite of taking the child out to bathe at a river (or sea) landing and teaching him to swim'.[11] It was performed throughout the Ayutthaya period but after Bangkok's establishment was observed only in 1812 and 1886. Photographs and drawings depict the enormous ceremonial platform as a *mondop* (a sacred square building with a multi-tiered roof) anchored in the Chao Phya River adjacent to the Grand Palace (Plate 50). The child, attended by servants and observed by its father, the king, would descend

50. The floating platform with its towering *mondop*, used in the royal Mahathitmongkon ceremony.

into the water through an opening in the centre of the platform while Brahman priests chanted ancient Sanskrit verse and various ceremonial objects—fish and shrimp made of gold, copper-gold, and silver foil—were placed in the water to ask the gods to bestow good fortune on the child.

All ordinary rites of passage in Buddhist Thailand utilize water. Until the middle of this century, a lock of hair was snipped from the head of every new-born baby. It was wrapped in a lotus leaf and floated on the river in a rite called Tak Phom Fai. For the Khuan Chak or tonsure ceremony signifying entry into adolescence, the topknot was severed, wrapped in a lotus leaf, and laid in a river as a symbolic floating away of misfortune. Today, this rite is regarded as an anachronism and it is rare to see a prepubescent child with a tonsure.

Symbolizing peace and happiness, water also blesses a newly wedded couple. In the past when marriages were arranged, monks performed the Sat Nam ceremony, sprinkling water on the couple and symbolically bringing them together. For the modern Rot Nam ritual, the couple kneel on cushions and rest their arms on a rail. A white ritual thread connecting their heads binds them together for eternity. Relatives and friends file past to pour perfumed lustral water from a conch over the couple's extended hands. Previously blessed by a monk, the water conveys wishes for the success of the marriage. The conch's mystical power derives from a Hindu belief that a demon once swallowed a sacred *mantra* and took refuge in a conch. To retrieve the *mantra*, the god Phra Narai squeezed the shell tightly, leaving his finger impressions along its rim.

At death, the body of the deceased is laid on a bier and covered with a cloth except for the right arm which is extended over a pan. Relatives and friends pour water from a conch over the deceased's hand to ask forgiveness for any deeds they might have committed against him or her. The body is then cremated and the ashes and bones are placed in a river in a Loy Aung-khan ceremony to float away sins and bring peace to the departed, in essence, 'burying' him or her in water.

Buddhist Rites

The three principal Buddhist holidays of Makha Pucha (Plate 51), Wisakha Pucha, and Asalaha Pucha, celebrating key events in the Buddha's life, are highlighted by a triple ambulation by monks and laity around the *wihan* (worship hall) of a Buddhist *wat* (monastery). As celebrants pass the *wihan*'s entrance door, the abbot gently flicks a bamboo whisk laden with holy water at them to bless them. Performed on the full moon night, these candle-light processions are likened to rivers of light flowing around the *wihan*.

51. A Makha Pucha candle-light procession is celebrated at Bangkok's Wat Benjamabophit.

It is incumbent on every Thai Buddhist male to enter the priesthood for a period of seven or more days. At the initial rite of his ordination, his head is shaved and the hair is wrapped in a lotus leaf and floated in the river. Then, in the monastery's *bot* (ordination hall), the abbot asks the novice a series of questions to determine his worthiness for the monkhood, one of which is: 'Are you a man?' According to legend, Phya Nak was transformed into a man and subsequently became a monk. One day, he accidentally reverted to his former self, and was caught and expelled. There is, however, a fear that Phya Nak might return in disguise. Phya Nak cannot lie, so were a novice to answer 'no' in response to the question 'Are you a man?', he would be unmasked as a *nak*. As a *nak* is ostensibly an animal, it cannot be ordained.

For the Kruat Nam, one of many Buddhist village rites, devotees pour water over their relatives or friends to bless them. In a related ritual, Song Nam Phra, water is poured over a monk to honour him. In some villages, women pour the water down a bamboo trough to ensure that they do not accidentally touch him and compromise his vows of chastity. After the morning sermon on *wan phra*, the weekly Buddhist holy day, worshippers along the lower Wang and Ping Rivers carry lustral water from the *wihan* to pour on the ground or a tree, usually a Pho tree (the species under which the Buddha meditated), in the *wat* compound. By extension, the act signifies pouring water on Mother Earth as an act of merit for the dead.

Buddhist monks rise at dawn to walk barefoot to receive food offerings from devotees who wait for them in front of their homes, a ritual known as Bintabaht. In the Central Plains, the monks paddle boats to receive offerings from water-side residents (Colour Plate 17).

Village Celebrations

Many Central Thailand water-related village rituals are founded on ancient animistic beliefs that were subsequently overlaid with Buddhist elements; others trace their origins directly to Buddhism itself. One of the most important rites honours the rice goddess, Mae Prosop. Villagers erect a square *chaleo*, a 'magic pentacle figure made of two interlaced bamboo triangles [*sic*]' or a fish basket, to indicate that a rice field is 'pregnant'. This talisman bars entry to boats or buffaloes which might damage the tender rice shoots. It also serves as a symbolic 'fence' in lieu of a real one.[12]

One ancient northern rite, the Poi Bang Fai, has survived only along the western border with Burma's Shan States. In a celebration similar to the Bun Bong Fai festival of Laos and north-eastern Thailand, home-made bamboo rockets are launched to induce the sky gods to impregnate the rain-producing clouds. The rite is thought to derive from a mythical Khmer king whose realm was plagued by a devastating drought. To alleviate the disaster, the king fired a rocket into the heavens to please the god Isuan (Shiva), who responded with a torrent of rain that drenched the parched earth. Also associated with Phya Nak, whose image often decorates these rockets, the rite reminds the gods of their duties to humans.

Monks are renowned as expert rocketeers. They build bamboo-tube rockets the thickness of a man's thigh. The tubes, often 10 metres tall, are then filled with gunpowder the monks have blended from secret *wat* formulae. While musicians beat long drums and cymbals to gain the gods' attention, the rocket's fuse is lit and the tenders scamper for cover, for good reason: rural rocket science is rudimentary and missiles can just as easily fly horizontally into the crowds as ascend to the heavens. The ceremony takes place in May, usually after the fields have been ploughed but before they have been planted.

A certain ritual, not particularly appreciated by its chief participant, is believed by villagers to produce rain. The Hae Nang Maeo (Cat Procession) originated in the North-east but on occasion is conducted in Central Thailand. Rarely performed—often only once a decade and only after Songkran when drought threatens—it begins late in the afternoon at the village *wat* where Buddhist priests chant prayers for rain as a female cat is placed in a cage. The cage is carried around the village, and at each house, villagers pour or toss water on the cat while inebriated marchers beat long drums and cymbals, dance, and sing to the cat (and by extension to the householders): 'Hail! Nang Maeo (female cat), give us rain, give us *nam mon* (consecrated water) to pour on the Nang Maeo's head. Give us cowrie (token money), give us rice, and give us a wager for carrying the cat.' Like the Poi Bang Fai procession, other songs have sexual connotations such as: 'The rain falls in four copious showers, a thunder bolt strikes a nun. Strip off her clothes and see the pudenda, the rain pours down heavily, pours down heavily.'[13] After circling the village, the procession returns to the *wat* where further prayers are chanted and the poor cat is released.

A form of sympathetic magic, the rite is founded on a cat's intense dislike of water. The more it howls, the more rain the gods will send, so it is continually soaked and a din is created to frighten it to cry louder. A female is used because she symbolizes fertility and abundance. As rain-making is an act of sex, it requires that the male sky god, through the medium of falling rain, pour 'semen' on to the 'womb' of Mother Earth which bears new life (i.e. tender rice shoots). If the rain fails to fall, the rite is repeated until it does.

The Pan Mek or 'cloud-shaping' ritual for rain-making is performed in Central Plains villages. Men mould clay figures of a man and woman embracing in a seated or supine position while chanting magical spells containing obscene words. The figures are placed by the side of a path or road to be seen by the sky god, who will be reminded of his relationship with the earth goddess and will 'fertilize' her with rain. A figure may also be carried in the Hae Nang Maeo procession.[14]

All of these rituals underscore an abiding belief that humans, in collusion with the spirits, can induce the gods to provide the necessities for a successful harvest. It also suggests that the gods are inconstant and not particularly intelligent or honest since they can be tricked or bribed to do human bidding. Their failure to respond with the desired results is attributed to their unpredictability or lack of proper attention when the rituals are being performed.

Tribal Rites

Given their dependence on rainfed agriculture, it is not surprising that northern tribes focus on the sky, not the streams in their rites. Propitiation rites for water spirits are conducted as part of a general ritual to ensure good harvests. At rice planting time, the Karens intone: 'Water Lord, Country Lord, Hill Lord, Mountain Lord, come down! Lord of Laykawkey village, Lord of Laykawkey stream, send us good rice, send us sparse weeds.'[15]

In the Mien (Yao) death ceremony, a 'boat' made of banana leaves is filled with ritualistic figures and objects. The priest shatters green reeds over the boat to convince the deceased that he has broken his bonds with the living world and passed into the realm of the dead. The reeds are then placed in the boat, which is pulled out the door and pushed into the 'sea', a jungled area far from the house, and burnt. The rite is based on the belief that 'spirits might well come back over land, but it is impossible for them to return over water'.[16]

The Lahu (Musur) tribe believes that old streams die and new ones begin to flow each New Year's morning. At dawn, children race to see who can reach the source of the stream first, and a woman from each family walks from house to house washing the hands and receiving the blessings of the house owner and his wife. Children honour their elders by pouring rivulets of water over them.

Perceptions of the River

From the foregoing, it is clear that water is perceived to contain spiritual properties and that it is regarded as a beneficent force by urban and rural Thais alike. When, however, water is regarded not as an independent element but as a flowing stream, it acquires a slightly more sinister connotation, one which extends to nature as a whole. Before considering the specifics of river spirits and related rituals, it is useful to examine this perception of nature in the raw because it has shaped the rationale for and mode of implementation of development programmes since the mid-twentieth century.

From even a superficial study, it is apparent that the Thais do not have the same regard for nature as that displayed by many other cultures. Their distrust is based in part on nature's unpredictability and seeming perverseness. This aversion is even more pronounced among city residents—a trait they share with most of the world's urban cultures. The distinction between rural and urban responses is important because while villagers are predisposed to accommodate a river's whims, urban peoples have an improper understanding of its dynamics and their response is to impose control over it, often with lamentable consequences. Anthropologist Srisakra Vallibhotama states it succinctly: 'In the Thai sense, the term *pa* [normally translated 'forest'] means "something that is not well ordered".'[17] The corollary of 'untidy' suggests that it must be reshaped in human terms and values.

Urban 'hubris' occurs in part because city residents regard nature as primitive and a reminder of the barbarism from which they have been saved by the humanizing effects of culture. Because urbanites feel only tenuous ties to an abstract nature, they emphasize civilizing the uncivilized. Since development decisions are made solely by Bangkok-based bureaucrats, this perception can impact in a negative manner on the environment.

To a Thai, rivers and water are perfect Buddhist metaphors. On the one hand, they represent tranquillity and happiness; on the other, they represent the mutable, unpredictable, and ephemeral nature of life, always changing character, always altering course. Drownings are not uncommon, and floods bring destruction and disease. Individual entities are not important in the totality of a river, but are absorbed into a larger whole. Like life, rivers have their own momentum; they pull one towards a distant goal, and it is useless to struggle against them.

Thus, Thais regard the river less as a benign abetter of daily life and more as a malevolent entity which must be appeased. The deeper and faster the water runs, the more potent its threat. At flood-tide, the river is an omnipotent force capable of killing those who venture into it. Today, as one watches village Thais in their daily intercourse with the river, it is apparent that while the river does not hold the terror it once did, it is still regarded as a mysterious, malicious power with which one must exercise extreme care and whose caprice cannot be predicted or underestimated. While they do not genuflect or honour the river each time they step into it, as do the Hindus of India, they accord it its rightful due and accept whatever punishment it metes out.

Some of these threats are very palpable. Until they were exterminated in the wilds in the 1970s, crocodiles were perceived as a menace to bathers and swimmers. A common greeting *rawang chorakhe* (beware the crocodiles) is still shouted at passing boatmen, or as a teasing warning to a departing guest. Today, monitor lizards lie on the sandy banks and, at the approach of a boat, scurry for shelter or leap into the water with a loud splash which Thais instinctively interpret as that of a crocodile.

Similarly, other underwater denizens contribute to a belief in the terrors of the river nether world. The *pla pakpao* (*tetra odon leiurus*, puffer-fish), which inhabits the mouth of the Chao Phya River, is thought to chew or rip its prey's extremities, one reason it is feared more by male than by female swimmers. With its barbed tail, its habit of feeding on the bottom, and its rough skin, the *pla krabaen* (*dasyatis bleekeri*, stingray) can frighten a wader, although it is not considered dangerous. Similarly, the long tendrils of underwater weeds, found primarily in the tributaries, can inspire panic if they wrap around a wader's legs, but do not otherwise pose a danger.

River Spirits

Because a river symbolizes the meandering vagueness of life, a watery element which defies predestination or control, it has acquired a host of deities and spirits. The spirits explain both the river's bounty and its seeming propensity for destroying land-based life-forms which venture or are swept into it. Thus, deities are propitiated to ensure that the river benefits rather than harms those living along it.

Paramount among the deities is Phra Mae Khongkha, goddess of the world's rivers who, as her Sanskrit name suggests, originated in India. After giving her name to the great river of the subcontinent, the Ganges, she probably arrived in South-East Asia with the initial trade contacts between India and the early Angkorian civilization of the sixth and seventh centuries. In Thailand, her name is generic, applied to all rivers, and she has derivatives in Burma, Laos, and Cambodia.

She is omnipotent, controlling the movements of the rivers and of the people who use them. Phya Anuman quotes nineteenth-century writer Sunthorn Phu in cautioning against insulting Mae Khongkha by word or deed:

In taking a bath at a riverside or stream, one should face the direction of the running water. The voiding of nature is prohibited. Do not face against the running water for one may accidentally be the victim of the black arts. After a bath, always pay respect to Ganga, the Water Goddess.[18]

More insidious is Phi Phrai, a malevolent spirit who not only kills people who invade her territory but who, on the weekly holy day, *wan phra*, and on Songkran, snatches the unwary and carries them to her underwater realm. The Phi Phrai is thought to be the restless spirit of a stillborn child or the ghost of a drowned adult. If another person comes to replace the victim, the ghost will be freed from its spell. Phi Phrais are found in all rivers, especially in the vicinity of whirlpools. In the North-east, Thais toss a cigarette and whiskey into a whirlpool as they pass to curry favour with Phi Phrai.

In some areas of the Central Plains, Phi Phrai is described as a woman with long hair who dwells at the river-bottom, usually in the vicinity of an overhanging bough. Although she has a golden comb and other golden ornaments, she is never satisfied with her possessions and can be heard wailing as she wanders through the night, entering houses to take whatever she finds. She watches waders, swimmers, and bathers from below and strikes them without warning, but is not considered malicious as she only wishes to take the humans to her underwater home to play with them and show them her treasures. She does not realize that they cannot breathe and will drown.

The logical explanation for the supposed touch of Phi Phrai on a person's submerged flesh is that one has brushed against a freshwater electric eel, entangling underwater weeds, or other submerged debris. The belief does, however, provide a cogent explanation for accidental drowning, the disappearance of objects, the felling of river-bank trees, and how gold of the type for which prospectors pan appears in a stream.

If rivers have their guardian spirits, so, too, do the boats which travel them. The boat goddess's name, Mae Ya Nang, translates as 'Mother, Grandmother, Lady', underscoring the female, nurturing, protective element of the spirit in the way that Western sailors refer to the sea and ships as 'she'. Originally the spirit of the tree from which the boat is made, Mae Ya Nang resides in the bow and bonnets of all vehicles including cars, buses, planes, and boats, bringing safety to the occupants and commercial success to the owner. On a boat, her fish basket hangs in the wheel-house, and the brightly coloured ribbons which signify her presence are tied around a cluster of flowers and incense on the bow. Passengers, especially women, are enjoined from stepping on the forward area for fear of trampling on the goddess's spirit. Should someone inadvertently tread on the bow, the owner will conduct special propitiation ceremonies to ask the goddess's forgiveness. Many owners of sampans and small boats forbid passengers to wear shoes while aboard.

Lower Chao Phya boatmen venerate a god that resides at a riverside shrine (Colour Plate 18) at Lante near Bang Sai, a few kilometres downstream from Ayutthaya. The shrine is a small wooden structure holding a *javet*, a statue of a male god in human shape. The *javet* is surrounded by small wooden sculptures of cows and buffaloes, animals said to be the god's servants. Sailors claim that passing boats stop of their own accord and refuse to proceed until Brahmanic rites are conducted. Even the engines of motorized boats have been known to die there. Boatmen do not waste time repairing the boat but go straight to the shrine to offer chickens, ducks, whiskey, flowers, candles, and incense. Incredible as it may sound, the engine usually restarts without difficulty. Wat Panancheng in Ayutthaya is also venerated by boatmen who believe that its Buddha images protect them.

Numerous other spirits explain natural phenomena. Phi Ha, the former name for cholera, is a spirit that brings pestilence. Early inhabitants seem to have recognized the connection between water pollution and disease but thought they could avoid illness by propitiating Phi Ha. They built shrines on the river-banks and took great care to ensure that the water near the shrine was always clean. It was believed that a polluter would immediately be stricken by the spirit and die a horrible death, and that all nearby riverine residents would sicken and die. The belief was a commonsense means of ensuring that villagers would have a clean source of water should the wells dry up. Would that the belief still prevailed and that shrines could be erected up and down the banks of the major rivers of Thailand to guarantee the water's potability.

Fishing was prohibited near the Phi Ha shrine in the belief that catching them was sacrilege. In one Central Thai folk-tale, a farmer tossed a net into the waters near a shrine and trapped a huge fish. Unable to pull it ashore, and wondering what kind of fish could weigh so much, he peered inside the net and found himself face to face with a ghost. Frightened, he dropped the net and scurried home. As he was bounding up the steps, the ghost caught up with him. The man toppled over backwards and was dead before he struck the ground.

Water-borne Processions

In a country where water is the principal highway for commerce and communications, it is natural that grand processions should be celebrated on rivers. Despite their setting, these festivals honour Buddha or monarchs but none honours the river itself. Moreover, their number has declined dramatically from the past when dozens of water processions—an important form of mass entertainment—could be seen the length of the river system.

From Ayutthayan times, the Chao Phya's chief water procession has been the Phraratcha Phithi Phra Yuha Yatra Cholamak, or Royal Barge Procession. Each October, the king travelled in his fleet of royal barges to present robes to the monks to mark the end of the three-month Buddhist Rains Retreat. The origins of this procession are unknown. Quaritch Wales suggests it was conducted in Burma and may have arrived there from other Buddhist lands. The ceremony was first seen in Thailand in Sukhothai. King Mongkut, in an article entitled 'Origin of *Wat* Visitations' and based on a reading of the Lanna Chronicles, concluded that the processions had begun in the reign of King Ramkamhaeng (1275–1317) and that they took place on the full moon nights of October and November. By the mid-Ayutthaya period, they were being conducted during the day.[19] Ancient royal barge processions were much grander than those of recent decades. A French Jesuit, Pere Guy Tachard, described the procession of 1686.

THE SPIRIT WORLD AND CULTURAL LIFE

Three and twenty Mandarins of the lowest Order of the Palace appeared first, every one in a Balon [barge] of State ... [they] advanced in file on two lines, and went along the side of the River. These were followed by fifty other balons of his Majesties Officers ... [which] had from thirty to sixty oars a piece.... After these came twenty balons more, bigger than the former, in the middle of which there was a very high seat, all gilt and spiring into a piramide [Colour Plate 19]; these were the balons of guards ... of which some had fourscore rowers.... Next after that long train of balons, the King appeared in his, raised upon a throne of a piramidal figure and extraordinary well gilt.... The King's balon was ... rowed by six score watermen.... Twenty balons ... followed the King's, and other sixteen half-painted and half-gilt, brought up the rear. We reckoned in all one hundred and fifty nine, of which the biggest were near six score foot long and hardly six foot over at the broadest place.[20]

Preparations for this ceremony would have required considerable organization. Jeremias Van Vliet, a Dutch merchant at Ayutthaya, recorded that the king required 20,000–30,000 men just to build and repair his boats. Even an ordinary royal procession to welcome a visiting dignitary might involve 25,000 persons.[21]

Adding to the spectacle was the presence of thousands of ordinary Siamese. Tachard wrote that in 1687, up to 200,000 people on 20,000 boats lined the banks of the river to watch the procession.[22] Since royal protocol forbade them to look upon the king or speak his name, they prostrated themselves, touching their faces to the ground until he had passed. Tachard was awed by its panoply of gilded oars and boats, brilliantly costumed mandarins and oarsmen (Colour Plate 20; Plate 52), stentorian chants that set the cadence, flagmen on the bows signalling manœuvres, and conch trumpets punctuating the magnificent din; he ranked it among the great celebrations of Asia. Certainly the European spectators were ecstatic in describing it, so different from anything they had seen in their own countries.

52. Royal Barge oarsmen dressed in Ayutthaya-period costumes.

The songs chanted during the Royal Barge Procession convey its pageantry as the majestic boats glide down the river. They are, however, paeans, not to the river, but to human achievement:

The King embarks upon the water
in his most magnificent barge.
Handsomely ornamented with 'King Kaew' [a crystal ball]
the movement of pliant paddles is beautiful to see.

Crowded together but preserving order,
each shaped in the semblance of a curious beast,
the vessels move with their flags flying,
making the water roar and foam.

Following Ayutthaya's fall, the royal barge fleet never again approached the huge numbers reported in the seventeenth century. In subsequent decades, it waxed and waned between 50 and 100 vessels. In the past century, the rising costs of maintenance and organization have severely limited the number of royal barge processions; in the past three decades, one was held in 1968 and another in 1982 to mark the bicentennial of Bangkok's establishment as the nation's capital and of the founding of the present Chakri dynasty, a period beginning in 1782 and known as the Rattanakosin era. The most recent procession took place in 1987 to celebrate the fifth cycle (sixtieth birthday) of King Bhumibol. The fleet numbered fifty-one vessels and the complement of Royal Thai Navy sailors trained to paddle them totalled 2,200 men.

The royal barges have also been employed when a grand spectacle was required to impress a foreign guest. When British envoy Sir John Bowring arrived in 1855 to induce Thailand to sign a Treaty of Amity with the United Kingdom, King Mongkut dispatched a complement of barges to Paknam to carry Bowring and his party upriver. The gesture had the desired effect: Bowring wrote effusively of his reception and journey to the Grand Palace.

On a smaller scale are the Buddhist waterway processions similar to those found throughout much of South-East Asia. The dates and durations vary but, in principle, one or more Buddha images, attended by monks, are placed on a large barge. As in a land-based procession, the image is paddled on a circuit of villages and shrines where devotees anoint it. This Chak Phra (Pulling the Buddha Image) was once celebrated in several riverside towns but has largely been abandoned. It persists on a Thonburi canal at Wat Nang Chi. Much reduced from the 200 boats with the hundreds of paddlers which participated several decades ago, it is celebrated as much for its curiosity and historical value as for its religious significance. On the full moon day of September, a barge loaded with Buddhist relics proceeds along the former course of the Chao Phya River, now Khlong Bangkok Yai and Khlong Bangkok Noi, where it is propitiated by canalside residents.

The Yon Bua or Lotus Showering ceremony is Mon in origin and is performed only on the Khlong Bangphli in Samut Prakan

THE SPIRIT WORLD AND CULTURAL LIFE

(Paknam) at the end of the Buddhist lent in October. Local legend says that 200 years ago, three sacred Buddha images floated down the Chao Phya, Bangpakong, and Tha Chin Rivers. Villagers tried in vain to pull the images to shore, but they seemed to have their own itineraries and resisted all efforts to dissuade them. The Luang Pho Sothon image eventually came to rest at Chachengsao; the Luang Pho Wat Ban Laem, in Samut Songkhram; and the Luang Pho To, in Khlong Bangphli at Samut Prakan. Unlike Chak Phra ceremonies, the Luang Pho To image does not leave the *wat* but by strewing lotuses on the waters of the Chao Phya, the Bangphli villagers honour it.

The Nan River serves as a medium for annual longboat races at Phichit and Nan each September and October. In the past, nearly every *wat* had a boat and team. Today, the craft generally measure 8 metres and are no wider than a pair of paddlers seated side by side. As its name, Phya Nak, suggests, the prow is shaped like the head of a *nak*, the deity which determines the crew's success. On the day before the race, team members chant ancient prayers to placate the *nak*, calling the spirit back from its wandering to protect the boat and its crew.

Rivers were also once prized as recreation areas. When Ayutthayan princes wanted to relax far from palace protocol, they *len thung* (played in the fields) (Plate 53) or *thon promat* (picnicked), rowing to islands in the river. Wat Mani Cholakhan on the island in the Lopburi River just north of the palace was a favourite site. Princes would pack only *kapi* (shrimp paste) and *nam phrik* (fish sauce with chillies), and once on the island, would pick water vegetables like *phak bung* and *santawa*, douse them with condiments, and eat them raw. These excursions continued into the Rattanakosin period; a celebrated photo depicts a casually

53. King Bhumibol in modern *len thung* on a canal. (Private collection)

dressed King Chulalongkorn relaxing in a boat, surrounded by his friends. Similarly, the *len pleng*, or song singing mentioned in connection with King Rama I's excavation of Bangkok's Khlong Mahanak, involved ordinary Thais pursuing a popular night-time amusement. These singing excursions provided an opportunity for courtship between young men and women.

The river was the setting for displays of fireworks designed to resemble fish. Gunpowder in the tail of the torpedo-shaped *pla chon* (serpent-head fish) made it leap and turn in the water as though it were alive. In some versions, the *pla chon* gave birth. Just as the large fish was winding down, it would burst and small *pla chon* fireworks would issue forth. The *pla krabaen* (stingray) was fitted with a palm leaf segment fashioned into a hydrofoil, causing the 'fish' to rear out of the water and skim along the surface. The *chorakhe* (crocodile) was also designed to skim, its blunt cork nose rising just above the water-line like its real-life inspiration. These firework displays were usually offered during the winter on moonless nights, the darkness enhancing the pyrotechnic effect.

The River and the Classical Arts

The formal arts, products of palace-based artisans, normally portray nature in a highly ornate, stylized form which suggests that natural objects were symbolized rather than depicted. In mural paintings, it is difficult to identify particular tree or plant species because they are presented only as decorative elements to illustrate the main story-line, not as realistic objects. Rivers in paintings are portrayed less as themselves than as highways for boats (Colour Plate 21). Thus, waterways appear as formalized, geometric motifs with imbricated waves and decoratively positioned mid-stream rocks suggesting only vague familiarity with rivers. In the murals in Wat Ratchabophit, Wat Pho, and Wat Patumawanaram in Bangkok, the rivers are mere backdrops for the vessels in the Royal Barge Ceremony.

Nature motifs in architecture are similarly depicted in a stylized form which reveals little about the natural organisms that served as the models. Oddly, in all the plastic arts, the essence of a river is best depicted by the sinuosity of line, in the near-liquid surfaces, and in the glass mosaics cladding monastery walls which shatter sunlight much like the sun glittering off a river's choppy surface does.

This distancing from nature permeates classical Thai literature, and reflects an urban orientation that regards the rivers simply as bodies of flowing water. One looks in vain in literature for depictions of the river as a spiritual entity, or for emotional response to it, even to something as wild and rugged as a rapid. Instead, Thai writers use nature as abstract representations of human emotions or endeavours. As Chamnongsri Lamsam Rutnin notes:

> While the natural environment would seem to be permanent, the Buddhist concepts of impermanence and the universal cycle of change are ever-present in the depths of Thai consciousness. This apparent contradiction of fact and philosophy results in a rarity of literary expression of purely esthetic wonderment and ecstasy concerning nature.... The role of nature in classical Thai literature is one of service to literary craft, creativity and expression.[23]

Wilaiwan Khanittanan underscores the point when she writes:

> ... the words 'nature' and 'natural environment' did not exist in the languages of the Tai peoples, so it would appear that Tai ideas about the natural environment were not expressed or given much consideration ... the words for forest (*pa dong* and *thuan*) have negative connotations.... In all Tai groups, any form of beauty, even natural beauty, was likened to something man-made, for instance, 'as valuable as polished gold'.[24]

The most popular form of poetry during the early decades of the Rattanakosin period was the *nirat*. Translated as 'separation', these poems were composed on long journeys, usually by boat. As Prince Damrong Rajanubhab noted, 'It is natural that [the poet] described things he saw along the way and related them to his moods and emotions.'[25] Most poets were unconcerned about the river but were, as the genre title suggests, filled with longing for a loved one left at home.

One exception is the great *nirat* writer Sunthorn Phu, who flourished in the court of Rama II. He embarked on long river journeys and commented unfavourably on the disappearance of the old ways. Even then, water and rivers were, for him, not living entities but metaphors for sensual and physical emotions which, in the formal, courtly tradition, could not be expressed in plain terms.[26] Similarly, in other forms of prose and poetry down to the present age, nature motifs appear as idealized, often fey, ornaments.

It may be argued that early Thais were engaged in a constant struggle with nature, seeing it as an enemy to be battled into submission so that crops could be planted. While the view is tenable from the perspective of early rural Thais, it is less comprehensible when applied to Thai artists, many of them urbanized but with roots in the land. Natural subjects are treated in the abstract and removed from human experience. There is no sense of oneness with nature, with humans as one of myriad components in it. There is little awe or appreciation of nature as a mystical entity. Instead, the writer stands apart, an observer who is superior to the nature which surrounds him. It is to the rural people that one must look for a sense of the importance of the rivers in daily life.

THE CHAO PHYA

The River in Folklore

Song and Stories

It is in rustic songs that one finds specific reference to rivers and their effects on the traveller or the riparian resident. For example, 'Sao Long Nan' translates as:

(Men)
Oh, admire the beauty
of the Ping, Wang, Yom, and Nan.
The water is clear and cool.
The wind blows along both banks.
So happy am I
when I float in a boat down the Nan

(Women)
Our Nan River holds many fish and crabs.
The scenery on both sides is beautiful.
Our Nan has its own soul
that we carry with us always.
Our minds are clear as the Nan.

River creatures figure in a number of folk-tales. The tale of Kraithong and his victory over the crocodile king Chalawan is a favourite throughout Thailand (Plate 54). Chalawan lives under the river in a magical cave. Each time he enters the cave, he is transformed into a human being. While travelling on land, he abducts the beautiful daughter of a merchant and takes her to his underwater realm. A poor but brave merchant named Kraithong follows him into his cave and in the ensuing battle captures him and rescues the daughter. The pleased merchant rewards Kraithong with wealth and his daughter's hand.

Proverbs

Dozens of Thai proverbs reveal an intimacy with the river and

54. Chalawan in concrete in the town of Phichit on the banks of the Nan River.

give the reader a sense of the life experienced along it. Many of these maxims date from the Sukhothai period when they were known as Phra Ruang's proverbs. A typical example would be: 'Do not put the boat lengthwise against a strong current', a corollary of 'go with the flow'.[27] In the nineteenth century, King Mongkut employed an ancient river proverb to explain his dilemma of being caught in a struggle between French and British colonial powers: 'It is for us to decide what we are going to do; whether to swim upriver to make friends with the crocodile or to swim out to sea and hang on to the whale.'[28]

Other river proverbs, created down the centuries and still recited, reveal a similar realistic but philosophical approach to life:

Mai hen nam tat krabok
mai hen krarok kong na mai
(He) doesn't see the river but cuts a bamboo tube [to store it].
(He) doesn't see the squirrel but cocks the crossbow.
An admonition to people who prepare for an event without being sure it is actually going to happen; thus, wasted or unnecessary effort.

Phut khlong pen long nam
Speak fluently like flowing water.
Pla kradi dai nam
The fish is happy underwater.
The fish that lies inconspicuous deep beneath the water's surface avoids being caught by the fisherman. Thus, it refers to someone so insignificant that he is overlooked by exploiters and thereby leads a happy life.

Pla mo tai phro pak
The mo fish dies because it opens its mouth.
The *pla mo* usually lies unobtrusively on the river-bottom. If, however, it opens its mouth, it releases bubbles that reveal its presence to a fisherman. This adage refers to someone who seals his doom by talking too much.

Mu mai phai
ya ao thao ra nam
The hand does not paddle,
Don't put your leg in the water.
If you are not going to help, at least do not hinder.

Nam khun hai rip tak
When the river rises, rush to throw your net.
Grab the opportunity when it presents itself.

Nam khun, pla kin mot
nam lot, mot kin pla
When the river rises, the fish eats the ant.
When the river falls, the ant eats the fish.
The wheel of fortune turns.

Nam ning lai luk
Still waters run deep.
Do not underestimate others.

Chai kwang muan maenam
Broad-minded like the river.
It also means generous.

Son chorakhe hai wai nam
To teach the crocodile how to swim.
A pointless task.

Mai lak pak len
A mooring pole not set deep in the mud [floats away].
Unsettled people are undependable.

Tang kan raw kap nam ka faa
As different as the river and the sky.
As different as night and day.

Children's Riddles

Like the proverbs passed from generation to generation, children's riddles and games reveal a familiarity with the river that is developed from an early age:

Malum, malam, den wan yang kham, mai hen roi = rua.
Malum, malam, walks all day until night, but leaves no trace = boat.

Arai euay, mee hang, mee pak, mee tha, kin pla, ben ahan = thak.
What has a tail, a mouth, eyes, and eats fish for dinner = a fish net.

Aarai euay, wela chai yon thing, wela mai chai ao pai kep wai thi hua = samaw.
What do you throw away when you use it, and put on the head when you keep it = anchor.

Children's Games

'PHONG PHANG'

Phong phang euay
Pla ta bot
Pla khao lot
Khao lod phong phang
The fish net
The fish is blind
The fish enters the net mouth
The mouth of the Phong Phang net.

Phong phang is a set of crossed bamboo poles from which a large net is suspended. The main pole is fixed to the river-bed in shallow water. In the game, a 'fish' is chosen by drawing straws. He/She is blindfolded and spun three times. Group members hold hands to form a circle, walking round and round the 'fish', while singing the 'Phong Phang' song. Then, they ask the 'fish' if he/she is dead or alive. If he answers 'dead', they must stand still while the 'fish' grabs someone from the circle and tries to guess who it is. If he answers 'alive', the children walk or run and the 'fish' grabs for them. If the 'fish' can identify his catch, then that person must take his place; if not, the 'fish' tries again.

'TA KHE TA KHONG/AI KHE AI KHONG'

Ai khe ai khong
Yu nai phrong mai sak
Ai khe fan hak
Kat khon mai khao
Little Crocodile
You live in the hollow of a teak tree
Your teeth break
So you cannot bite people.

Children draw straws to select the *ai khe* (crocodile). The *ai khe* stands in an imaginary river while the rest of the children divide into two groups to stand on the banks on either side. They sing the above song to mock the crocodile. They must 'swim' (i.e. run) across the river while singing the crocodile song. Whomever *ai khe* catches becomes the next *ai khe*. It can also be played in a canal or river.

'CHAM CHI' (A FINGER-COUNTING GAME)

Cham chi ma khua po
Krato na waen
Phai rua ok an
Sao sao, num num
Ap nam tha mai
Ap nam tha wat
Ao paeng nai phad
Ao krachok nai song
Yiam yiam, mong mong
Nok khun rong wu
Cham Chi, the vegetable
There's a crack in the glass
We paddle a boat very hard.
Girls and boys,
Bathe at which dock?
Bathe at the temple dock.
Where did you get the powder?
Where did you get the mirror to see yourself?
Look look, see see,
The mynah says 'Woo'.

Children sit in a circle around a 'counter', and put their hands on the floor. As he sings each syllable, the 'counter' points at each finger of the players in sequence. When the last syllable falls on a finger, that finger must be tucked away. Then the song and the counting begin anew and continue until only one player is left.

Armed with an understanding of the rural and urban perspectives on rivers, it is easier to understand the grand schemes which have been initiated to harness them, projects which have proven to be the engineering equivalent of 'placing one's boat lengthwise against a strong current'.

1. Wilaiwan Khanittanan, 'The Order of the Natural World as Recorded in Tai Languages', in *Culture and Environment in Thailand: A Symposium Sponsored by the Siam Society*, Bangkok: Siam Society, 1989, p. 236.

2. Ibid., p. 235.

3. Paul Lewis and Elaine Lewis, *Peoples of the Golden Triangle*, London: Thames and Hudson, 1984, p. 254.

4. S. Singaravelu, 'The Legend of the Naga-princess in South India and Southeast Asia', Paper presented to the 25th Session of the All-India Oriental Conference, Calcutta, 29–31 October 1969; reprinted in Tej Bunnag and Michael Smithies (eds.), *In Memoriam: Phya Anuman Rajadhon. Contributions in Memory of the Late President of the Siam Society*, Bangkok: Siam Society, 1970, p. 10.

5. T. E. Cirlot, *A Dictionary of Symbols*, New York: Barnes and Noble, 1993, p. 286.

6. Phya Anuman Rajadhon, *Essays on Thai Folklore*, Bangkok: Social Science Association Press of Thailand, 1968, pp. 41–4.

7. Sommai Premchit and Amphay Dore, *The Lan Na Twelve-month Traditions*, Chiang Mai: Toyota Foundation, 1992, p. 68.

8. Sumet Jumsai, *Naga: Cultural Origins in Siam and the West Pacific*, Singapore: Oxford University Press, 1988, p. 33.

9. Ibid., p. 39.

10. Sommai and Amphay, *The Lan Na Twelve-month Traditions*, p. 215.

11. G. E. Gerini, *Chulakantamangala: The Tonsure Ceremony as Performed in Siam*, Bangkok, 1895; reprinted Bangkok: Siam Society, 1976, p. 2.

12. Anuman, *Essays on Thai Folklore*, p. 339.

13. Ibid., pp. 196–7.

14. Ibid., pp. 199–200.

15. Lewis and Lewis, *Peoples of the Golden Triangle*, p. 97.

16. Ibid., p. 164.

17. Srisakra Vallibhotama, 'Comment during Roundtable Discussion', in *Culture and Environment in Thailand: A Symposium Sponsored by the Siam Society*, Bangkok: Siam Society, 1989, p. 269.

18. Anuman, *Essays on Thai Folklore*, p. 71, quoting from the 'Swasdi Raksa', Sunthorn Phu's homilies for princes.

19. H. G. Quaritch Wales, *Siamese State Ceremonies*, London: Bernard Quaritch, 1931, p. 210.

20. Guy Tachard, *Voyage to Siam*, London, 1688; reprinted Bangkok: White Orchid Press, 1981, pp. 187–9.

21. L. F. Van Ravesswaay, 'Jeremias Van Vliet, Description of the Kingdom of Siam', *Journal of the Siam Society*, 1910, pp. 25–6.

22. Tachard, *Voyage to Siam*, p. 190.

23. Chamnongsri Lamsam Rutini, 'Nature in the Service of Literature', in *Culture and Environment in Thailand, A Symposium Sponsored by the Siam Society*, Bangkok: Siam Society, 1989, pp. 234, 243.

24. Wilaiwan, 'The Order of the Natural World as Recorded in Tai Languages', pp. 233–4.

25. Manas Chitakasem, 'The Emergence and Development of the Nirat Genre in Thai Poetry', *Journal of the Siam Society*, Vol. 60, Pt. 2, July 1972, p. 138, quoting Prince Damrong Rajanubhab in his preface to the Prachum Nirat Sunthorn Phu, Bangkok, 1922, p. 2.

26. Chamnongsri, 'Nature in the Service of Literature', pp. 250–1.

27. Sumet, *Naga*, p. 145.

28. MR Seni Pramoj and MR Kukrit Pramoj, 'Letter to Phraya Suriyawongse Vayavadhana, Siamese Ambassador to Paris, 1865', in *A King of Siam Speaks*, Bangkok: Siam Society, 1987, pp. 176–7.

6 Taming the River

It is a Thai tradition to improve upon nature with craft.[1]

SURROUNDED as they were by water, it was inevitable that the Thais would excel as hydraulic engineers, modifying rivers and digging canals to serve development needs. The thousands of kilometres of tiny waterways that thread through Ayutthaya's, and later Bangkok's, suburbs, paralleling and intersecting each other like rural roads, suggest a culture of inveterate, indefatigable excavators. On the Chao Phya, the scope of their work expanded from the *khlong lat* short-cut canals of the Ayutthaya period to the grand *chuam maenam* canals of the early nineteenth century dug to convey soldiers or goods, drain low areas, and facilitate provincial administration. By the 1830s, the cessation of wars with neighbouring countries and the rise of a new economic and political dynamism dictated a need to improve internal communications by excavating metropolitan waterways and rural canals to open up land for cultivation, to transport produce from the farms to Bangkok and other urban centres, and to send goods to provincial towns.

Until the nineteenth century, construction of short cuts and auxiliary canals had not significantly altered the river's basic structure. That changed after the 1850s, beginning with the 1857 alteration of the Chao Phya's course to flow through Ayutthaya (see Chapter 1). The impetus for the next phase of canal construction was the attractive prices offered in world sugar markets. Introduced into Thailand in 1809, sugar was identified by the Bowring Treaty of 1855 as an export product with enormous potential. Between 1857 and 1868, four new *chuam maenam* canals (to connect one river to another) (Map 11), initiated and financed primarily by royalty, opened the western Central Plains to sugar-cane cultivation. In 1860, Khlong Mahasawat (17.1 kilometres long, 14.0 metres wide, and 3.0 metres deep) was dug to link Bangkok with the Tha Chin River. Financiers who provided money or labour for its excavation were awarded cultivable land along its banks. The main beneficiaries were nobles and high-ranking officers, the only Thais other than the royal family with sufficient funds to underwrite large-scale projects. Khlong Chedi Pucha, an extension of Mahasawat, began at Tha Nam on the Tha Chin River and ran 11.225 kilometres to Wat Phra Ngam in Nakhon Pathom. Most of its construction costs were borne by King Mongkut from sales of assets confiscated from Thao Thep Akon Talat, a prince who

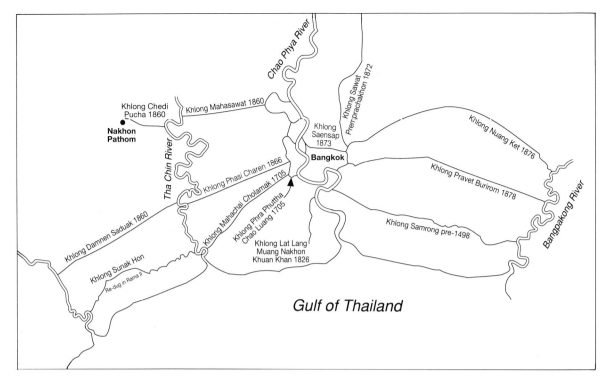

Map 11. Lower basin canals.

had acquired a fortune by manipulating markets during the reign of Rama III. The land on either side of the canal was granted to Mongkut's sons and daughters, who were allowed to farm it with corvée labour.

Another set of canals was dug south of, and parallel to, Mahasawat–Chedi Pucha. Damnen Saduak, 21 kilometres long, linked the Tha Chin River at Bang Yang to the Mae Klong River at Bang Nok Kwang. Dug between 1860 and 1868, it was financed primarily by sugar tax revenues originally intended for construction of a royal palace at Phetchaburi.[2] Khlong Phasi Charen (fruit of tax) (1866–72) was financed by the 'opium tax farmer', a private citizen who had purchased a government concession to collect duties on opium production. The canal ran 15.5 kilometres from Khlong Bangkok Yai at Wat Pak Nam to Ban Don Kai Di in Samut Sakhon province. Fourteen metres wide, it quickly became a major trading route between towns along the Tha Chin River and Bangkok.

Portents of the Future

In the 1870s, King Chulalongkorn began modernizing Thailand's administration and economy. The river was now regarded, not as an untameable force, but as an engine for development. His reign saw the most ambitious canal-digging initiative in Thai history. Concentrated in the lower Chao Phya River Valley, the work programme comprised two phases to serve two purposes. The first phase (1870–85) created canals to transport sugar-cane between plantations in the west and Bangkok. The second (1886–1910) opened forested land north-east of Bangkok to rice cultivation.

Mongkut's canals had enabled western basin plantation owners to float their harvests to Bangkok. This new ease of transport, encouraged them to expand their cultivation areas, clearing and planting cane on land further west. Unfortunately, sugar's profitability was linked to world markets, the first time the Thai economy had been so exposed to foreign price fluctuations. Sugar's high prices encouraged new countries to compete, and supply soon outstripped demand. When Indonesia began producing sugar in the 1870s, the world price plummeted. This set-back, coupled with floods and droughts in Nakhon Chaisi, served to scuttle Thai sugar entrepreneurs, not the first time in Thai history that farmers would be betrayed by the vagaries of the market-place.

When sugar prices failed to recover, the government shifted its focus to rice. Originally conceived to serve sugar-cane interests, Khlong Sawat Premprachakon became a conduit aiding rice farmers. Completed in 1872, the 31.8-kilometre canal connecting Khlong Padung Krung Kasem and Khlong Kho via Don Muang was the first to be surveyed with theodolites. Canal financiers were awarded deeds to land on both canal banks in the hope they would cultivate it. When it became apparent that many new owners had no interest in putting the land to productive use, this approach was abandoned.

To aid new cultivators and to facilitate communication with eastern towns, Khlong Nakhon Nuangket was dug in 1876–7. The 63.25-kilometre-long, 12-metre-wide canal from Khlong Sansaep to Khlong Tha Kai shortened the distance between Bangkok and Chachengsao. To overcome the problem of absentee landlords, Chulalongkorn decreed that a would-be cultivator had to plant a crop within three years of receiving the title deed or forfeit ownership.[3] The 28.7-kilometre Khlong Pravet Burirom (1878–80) also shortened the route between Bangkok and Chachengsao, offering a southern alternative to Khlong Nakhon Nuangket. Beginning at Khlong Prakhanong, it ended at the western entrance to the 25-kilometre-long Khlong Tha Thua, which intersected the Bangpakong River 16 river kilometres below Chachengsao. Four secondary arterials provided access to new farmland. It is recorded that many of the Chinese who dug the canals were paid in opium.[4]

After 1880, canal construction waxed and waned according to the rice price. During 1881–5, rice prices fell sharply and canal construction ceased. In 1886, the price recovered, and in 1887, the second period of canal digging began, and a new form of financing was adopted. Contracts and concessions were sold to royalty, nobles, high officials, and Chinese businessmen who were allowed to dispose of the land as they chose.

A New Approach

Beginning in the 1870s and covering a period of twenty years, Rama V issued a series of royal decrees which dismantled and ultimately abolished the system of corvée labour and debt slavery. A flood of Chinese immigrants ensured a plentiful, relatively inexpensive supply of labour for canal dredging. Many administrators considered them to be more motivated and productive than Thai corvée labourers, who worked only to satisfy an antiquated system. The eradication of serfdom created a large manpower pool to clear and cultivate vast new tracts of forest land in the lower valley, an endeavour which contributed to the expansion of Thailand's role as a major rice exporter. These initiatives shaped the next and most ambitious phase of canal construction, the Rangsit Project.

Initiated in 1890, the Rangsit Project, Thailand's first comprehensive irrigation programme, was designed to expand rice cultivation into virgin lower basin lands. Two grids totalling 1600 kilometres of waterways were to be dug over a period of twenty-five years. The first was concentrated in the Rangsit area 20 kilometres north-east of Bangkok; the second phase would open land on the western side of the Chao Phya near Suphan Buri. Each canal differed from traditional channels in having a set of gates at the junctions with rivers or larger canals. Each gate could be

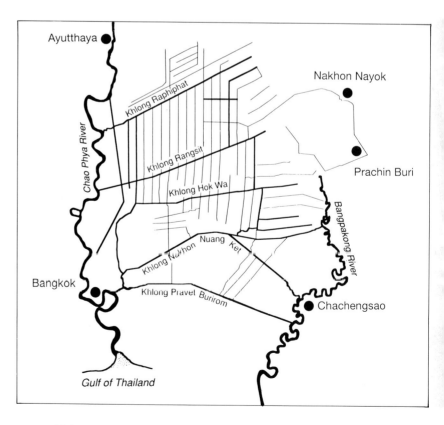

Map 12. The Rangsit Project.

TAMING THE RIVER

winched up or down to control the level and flow of water.

As the complexity and scope of the project required Western technology and machinery, the Thai government commissioned a specially formed Thai and Italian joint venture called the Siam Land, Canals, and Irrigation Company (SLCIC) to carry out the work. It thereby gave private interests a virtual monopoly on canal construction and land development throughout the kingdom for the next twenty-five years. The company would finance the operation with its own funds and, in return, be granted the right to cultivate or sell the land the project encompassed.

The basic design was straightforward. A main 12-metre-wide, 3-metre-deep east–west canal (Khlong Rangsit) was dug as a 54.8-kilometre spine from the Chao Phya River to Ongkharak where it joined a series of canals that carried on to Nakhon Nayok (Map 12). Extending at right angles north and south of the Rangsit were the ribs: 42 sub-canals 6–10 metres wide and 1.5–2.5 metres deep. The 13 longest sub-canals were spaced 2.5 kilometres apart and ran north 21.5 kilometres from Khlong Rangsit to a parallel canal, Khlong Raphiphat. A second set of canals was dug from Khlong Rangsit to Khlong Hok Wa, 12 kilometres south of and parallel to Khlong Rangsit. Other canals extended south from Khlong Hok Wa or ran east–west to connect some of the north–south canals. Engineers used mechanical shovels mounted on railcars or on barges; in difficult areas, they employed Chinese labourers. By 1900, nearly 80 000 hectares had been cleared; by the completion of the company's activities, 200 000–240 000 hectares had been brought under cultivation.[5]

55. One of the watergates installed in the Rangsit Project. This is between Khlong 13 and Khlong Hok Wa.

By 1900, however, the government was having misgivings about this approach. It was concerned that a private company was determining the pattern of agricultural development and settlement in the lower basin (Plate 55). In addition, by controlling the water flowing into the Rangsit area, the project denied water to farmers tilling fields just to the south. Most distressing of all, the government found it was unable to control the quality of construction. Many of the new canals were too narrow for their depth and were silting up, and the SLCIC was unwilling to redredge them. After completing only 835 kilometres of the system, the SLCIC was informed that its commission agreement was rescinded, and the government sought a new arrangement over which it could exert greater control.

1902: Watershed Year

The years 1902 and 1950 mark two watersheds in the river's history. Prior to this time, the river had shaped the people who adapted their lifestyles, boats, and homes to its moods. After 1902, humans began to shape the river to their needs. A new imperative—speed—in the form of railways and cars, provided the initial reason to turn away from the slow, winding river.

In 1899, Charles Rivett-Carnac, Financial Adviser to the Thai government, had stressed the importance of irrigation canals in the development of Thailand. It was they, he claimed, not railways or roads, that were the vehicles of progress for the journey into Thailand's future. His continual pleadings went unheeded, however, because his listeners' attention was distracted by a pivotal event far from Bangkok. On 25 July 1902, Shan rebels attacked government offices in the northern town of Phrae and killed twenty provincial officials. Troops from Bangkok reached Phrae by boat after the renegades were already well entrenched, making it difficult to quell the rebellion. The delay was blamed on the inability of river transport to move troops speedily to a trouble spot. In 1887, engineers had surveyed a route for a northern railway but Chulalongkorn had not approved its construction, concentrating instead on building a 264-kilometre railway to Khorat that was completed in 1900. Deciding he could delay no longer, he issued instructions in August 1902 to begin construction of the Chiang Mai line.[6]

On 13 June 1902, six weeks before the Phrae uprising, the government had established the Canal Department. In the same month, J. Homan Van der Heide, a Dutch irrigation engineer with extensive experience in Java, arrived to take up a government commission to design a comprehensive irrigation, drainage, and flood control scheme for the lower Chao Phya basin. After considerable research, he proposed building a concrete diversion dam across the Chao Phya at Chainat to create a reservoir. Canals would be dug on either side of the Chao Phya to run from the dam to Ayutthaya. From them, ancillary irrigation canals would run east and west into the fields. The twelve-year scheme would irrigate more than half of Thailand's existing cultivated land. He suggested that the B47 million construction costs be recouped by charging farmers fees for using the water and by selling newly irrigated land.

When briefing Van der Heide, government officials had neglected to provide him with budget guidelines, so were shocked by his cost estimate. Financially constrained by commitment to the northern railway and unwilling to spread canal construction costs over several years, the government asked Van der Heide to devise a less expensive alternative. Meanwhile, in 1903, the government decided to shore up existing canals and construct watergates on Damnen Saduak, Phasi Charen, Pravet Burirom, Saensap, and Samrong Canals to improve the year-round flow of water.

In 1906, Van der Heide presented a revised plan, essentially a reworking of the 1903 scheme but with several components excised from it. The four-year project would cost B24 million. The government still found it too expensive and asked Van der Heide for yet another proposal. In 1908, he designed a dam and canal system on the Pasak River to improve irrigation in the Rangsit region. Even this reduced work programme was rejected on the grounds that its benefits were too meagre to justify its costs. In August, the

Thai government's financial adviser, W. K. F. Williamson, argued: 'To my mind it has not yet been satisfactorily shown that new irrigation works are required in Siam, except as feeders to already existing systems, owing to the want of sufficiently dense population.'[7]

In 1909, rice and spice crops in fields adjacent to Khlong Damnen Saduak and Phasi Charen were destroyed by floods and Van der Heide was accused of incompetence in misreading Thai geology, in failing to carry out soil tests, and in constructing faulty flood control gates that buckled in the soft earth. In 1909, the government abandoned all plans for large-scale irrigation and Van der Heide left Thailand. By 1912, work was under way on the northern railway which would parallel two rivers and cross three, ultimately supplanting them as the prime transport links between north and south. In April 1912, to underscore its disinterest in waterways, the government transferred the Canal Department from the Ministry of Agriculture to the Ministry of Communications and renamed it the Krom Thang (Department of Ways).

On ascending the throne, King Rama VI appointed a new council of ministers. In 1913, his Minister of Agriculture, Prince Raturidirekrit, argued forcefully for a comprehensive irrigation scheme to enable Thailand to compete successfully in the world rice trade. He urged the government to abandon its policy of digging canals solely to open up new land, and to concentrate on improving yields by providing irrigation water. A disastrous flood in 1911 and droughts in 1912 and 1913 undoubtedly bolstered his case. Initially, he encountered opposition from railway proponents, but he prevailed and a new foreign canal engineer was engaged.

Thomas Ward, a Briton, was given a budget of £1.75 million, and in 1915, he unrolled his blueprints. He rejected Van der Heide's Chainat Dam proposal as too costly, contending it would open up more land than there were farmers to cultivate it. Instead, he suggested that the Pasak River be dammed to serve the Rangsit area, and that several small canal projects be undertaken to support cultivation in the vicinity of Suphan Buri. Funds would be transferred from the Chiang Mai railway project, which had encountered logistical problems while trying to cut through the northern mountains. The Minister of Finance supported the plan, arguing that with World War I raging, it was difficult to secure European rail engineers or materials. Any delay in completing the irrigation system would give neighbouring nations the advantage in the rice trade and, in time of war, it would be difficult to earn foreign revenue from other sectors of the Thai economy. Most importantly, the railway required foreign loans; financing for the irrigation system could be found within Thailand. In the end, economic considerations won out and the plan was approved.[8]

Ward wanted to begin with what came to be called the Suphan Project, which would have irrigated an area the size of the Rangsit Project but on the western side of the river. He was told, however, to concentrate instead on the Pasak scheme. A 1927 Royal Irrigation Department (RID) report suggests that in reaching that decision, personal interests had prevailed over public welfare. Influential Rangsit landowners realized that newly opened land in the Suphan Buri area would draw their cultivators away, depriving them of rents; the Pasak scheme, just north of Rangsit, would benefit them directly. Having won the war, the Finance Minister was prepared to concede a battle; the Pasak Dam (later renamed the Rama VII Dam) project was completed in 1924 (Plate 56). Even floods in 1918–19 and droughts in 1919–20 and 1920–1—the first drought leading to crop failure so severe that the government banned food exports to ensure domestic supply—could not sway opponents to support additional irrigation schemes. Work began on the Suphan Project but budget constraints and the subsequent world depression halted the work in the 1930s. Between 1915 and 1940, only B47 million was expended on irrigation works.

Economist James C. Ingram has postulated four possible reasons for the government's reluctance to embark on a comprehensive irrigation programme. First, the project was not self-financing and farmers would baulk at paying for water. Second, the project would not achieve dramatic results as it would had Thailand been an arid country like Egypt. Third, Thailand lacked sufficient farmers to cultivate the vast lands that would be opened up. Fourth, other infrastructural and strategic projects like railways had more claim to the limited funding available at the time.[9] Given the reluctance shown by the government down to the present day to undertake costly, long-term infrastructural projects—opting instead for stopgap measures to counter catastrophe—Ingram's contention has merit.

56. The Rama VII Dam on the Pasak River, the first major dam to be built in Thailand.

Water use fees might have financed a major portion of canal construction costs but imposing a tariff on farmers was considered a highly political issue as it would be when raised in 1948 with regard to the Greater Chao Phya River Project and again in the 1980s. After deducting the costs required to administer the system, gauge the amount of water used, and collect the fees, it is arguable that the ultimate revenues would have been too meagre to justify creating a user fee scheme. Even a single-time charge assessed per season would have posed an administrative problem given the poor communications network of the time. The contention that for centuries northern farmers had been paying pre-set *muang fai* fees at the beginning of each planting season was rejected as unrelated to the situation.

New Directions

In 1927, the Krom Thang was upgraded to the Chonlaprathan (RID) and was charged with improving northern irrigation facilities. The RID began its work in the North by constructing the Mae Faek Dam. Built on the Ping River, north-east of Mae Taeng, the concreted stone water storage dam was completed in 1935. Emboldened by its success, the RID then built permanent barrages on the Mae Ping Kao, Mae Kuang, and Mae Taeng Rivers. Together, the four dams irrigated 51 700 hectares of farmland or one-third of the Chiang Mai–Lamphun basin. On completing the Mae Ping Kao barrage in 1939, the authorities abolished all wooden weirs along the Ping River, replacing them with control boxes to regulate the flow of water into lateral canals for distribution to the fields. These four irrigation projects changed northern planting patterns, enabling many farmers to double crop and to plant higher-yielding rice varieties. As an additional benefit, the city of Chiang Mai suffered no dry-season water shortages after the Mae Taeng project was completed.[10]

After 1940, the RID focused on eliminating the villagers' cost and labour inputs of weir upkeep—expenses unacceptable from the standpoint of efficiency and economy. As they aged, the weirs shifted and their efficiency decreased. While villagers reckoned a weir's life span at 300 years, interviews with village elders revealed that even the best-built systems were viable for only 100 years before they had to be completely rebuilt. There is little doubt that labour and material costs had become unacceptably high. For example, after a 1975 flood, it took 8,000 freshly cut trees to replace one weir on the Mae Chaem, a Ping tributary.[11] The RID began to replace traditional wood-and-stone weirs with dams of concrete or loose stones (*hin ko*). As construction of these dams was usually beyond the budgets or technical expertise of the villagers, the RID provided financing, materials, heavy equipment, and engineers to design and direct their construction. With the Mae Chaem weir, it could be argued that by their insistence on concrete dams, the RID was more environmentally friendly than the villagers since it eliminated the need to cut watershed trees.

Under the Fourth National Development Plan (1977–81), the Mae Ngat (completed in 1985) and Mae Kuang (completed in 1987) Dams were built on Ping tributaries north of Chiang Mai. The dams increased the North's irrigation area to 72 470 hectares, or nearly half the Ping River basin, an amount comprising 2.3 per cent of Thailand's total irrigated land. Unlike the four dams built in the 1930s, these two earth-filled dams were equipped for flood control and hydroelectric power generation. With storage capacities of 265 and 363 million cubic metres respectively, they were also considerably larger than their predecessors.

Eroding Muang Fai Hegemony

The RID's involvement in northern irrigation development had far-reaching ramifications on the structure of *muang fai* administration. For seven or more centuries, the upper North's irrigation needs had been admirably served by a system which not only controlled the rivers and streams but defined the social fabric of thousands of villages, welding communities together to share resources to ensure mutual prosperity.

As part of its northern irrigation project, the RID in 1934 promulgated the People's Irrigation Act—subsequently revised in 1935, 1937, 1939, 1980, 1983, and 1993—which was based on an amalgam of older village *muang fai* contracts. In addition to the village *muang fai* organizations, the Act covered a number of private *muang fai* systems which had been built by northern royal families and wealthy landowners using corvée labour. Farmers willing to pay the royal owners a portion of their crops had been permitted to use the water. The 1939 version of the People's Irrigation Act required private *muang fai* owners to register their systems, effectively bringing them under RID administration.

The 1939 Act detailed rights and responsibilities of *muang fai* associations. It also replaced many older organizations with water user groups and associations, and government officials became members of village committees. Many farmers were unaffected by the Act because, in many instances, they were unaware of the law's existence. Others, finding the Act unfathomable and filled with loopholes, continued to operate according to the old *muang fai* agreements, perpetuating a tradition of local autonomy and resistance to RID water management directives. In 1942, the government passed the National Irrigation Act, a comprehensive guideline for irrigation development and administration throughout the country, and the People's Irrigation Act was incorporated into it. In theory, this expanded Act reflected a unified national approach to water management. In succeeding years, it would govern provision of water to farms, cities, and industries, and would challenge northern farmers' claims to primacy in extracting unlimited volumes of water from the streams flowing past their fields.

By the 1960s, however, there were signs that while it still functioned in many areas, the traditional *muang fai* system was under considerable strain. New land adjacent to the older networks was being cleared and planted, but the system was too rigid to accommodate it. The lower valleys were being invaded by hilltribe farmers and Central Plains settlers who threatened the tight-knit *muang fai* social structures as well as blurred the ancient boundaries between systems. In addition, the old taboos no longer curtailed over-exploitation of resources as they once had. These factors combined to weaken the contracts governing *muang fai* organizations. Reduced penalties suggest the decreased importance of the system and the difficulty of enforcing its rules. From the beatings and outright executions of an earlier era, punishment had been reduced to paltry fines. By 1962, the *muang fai* agreement for Sankampaeng, for example, would state: 'Anybody who stole water, enlarged the channel, and destroyed dividing structures must pay 10 baht for each farm turnout.'[12]

The village organization that had contributed significantly to the success of the system was also changing. Formerly, village populations had been stable and the force of tradition had been strong. Villagers might take periodic employment in the cities but they continued to regard themselves as members of their community, beholden to its rules. Now, village cohesion was weakening, and with it, the willingness to co-operate. In several instances, outsiders were hired to repair the weirs, supplanting the older tradition of local work parties as part of a collective, co-operative village effort. Most distressing to *muang fai* administrators, the new settlers refused to pay for the water; old farmers, seeing the newcomers receiving free water, became reluctant to pay.

Competition for dwindling water supplies erupted into open clashes. As Jack Potter notes in his study of the *muang fai* system in Saraphi, community conflicts had previously been resolved by calling upon the bonds of kinship and the authority of the village headman. Now, whenever several villages with their own *muang fai* systems vied for different portions of the same stream or when supplies were scarce during the dry season, the potential for conflict escalated. The perception by downstream villagers that an upstream community was channelling more than its fair share of water into its fields often led them to destroy the weirs of their upstream neighbours.[13]

The increased competition for water was also exacerbated by the rising population within individual valleys. Logging for village needs and commercial sale was destroying watersheds, and the resultant erosion was reducing soil fertility. Chemical fertilizers, pesticides, and herbicides reduced water quality and poisoned the land. Furthermore, the government was urging farmers to plant irrigated rice and second crops, just as two decades later it would encourage farm corporations to clear northern hillsides and plant cabbages for export.

Many of the *muang fai* systems had simply grown too large for efficient administration. At the same time, farmers were becoming increasingly sophisticated and politically aware, less willing to think in terms of the community or to bow to a village leader or to spirits. This combination of factors meant that many villagers were amenable to ceding water management matters to an overlord prepared to take on the financial responsibility for the system. These farmers also preferred the labour reduction that a system of concrete or loose stone weirs and channels would provide. Such an authority could also intercede in disputes. In the past, Chiang Mai's royal government had provided this function, albeit on a limited scale. By the 1960s, the only state-level organization with such a capability was the RID.

Most *muang fai* systems continued to function unchanged through the 1950s. In 1962, in a departure from tradition, farmers in the People's Irrigation Systems were charged a fee for water use. The initial fee was 2.0 litres of rice per farmer with an additional 5.0 litres of rice if he cultivated his own land, and 15.6 litres of rice per hectare from tenant farmers. In addition, water users paid the *hua na fai* B31.25 per hectare every dry season to cover weir maintenance.[14] A similar scheme was introduced in the Mae Taeng project in 1970 but all fees were paid in cash. Fees were assessed on landholdings and did not limit the farmer in the amount of water he used except as dictated by the *hua na fai*.

Through the 1970s, new provincial roads and trucks provided linkages between village and city, enabling farmers to supply produce to distant buyers. It also allowed government officials to assume a greater role in local affairs. A village headman was now answerable to a district office and to a larger development scheme than his immediate constituency. Water allocations were determined at weekly district agency meetings and communicated to village headmen by radio. Local *muang fai* organizations were administered, not by villagers but by government authorities.

Such government intervention further weakened traditional *muang fai* administration. In 1983, Richard Lando noted, 'As the importance of the government officials in the irrigation committee increased, the prestige and power of the office of irrigation headman has decreased.' Whereas in the past, villagers had actively sought the position of *muang fai* administrator, seeing it as a stepping stone to the *kamnan* (subdistrict headman) position, it was now shunned, in part because the *kamnan* had become a quasi-government official, with irrigation matters as part of his official duties.[15] In addition, the *hua na fai* lost his authority to resolve disputes or to conscript labour for repair work, ceding his decision-making rights to the district official.

As is common in other areas of rural life, when the impetus for an organization no longer comes from within the community itself, the motivation to maintain it is reduced and the organization's effectiveness is hampered. It also means that water usage is determined by urban professionals, often with insufficient comprehension of local conditions, needs, or politics. As a result, governing policies bear little relationship to the realities of farm life.

The organization most adversely affected by the imposition of external administration has been the Water Users' Association. Farmers are unwilling to pay fees for water or weir maintenance. They know they will receive water regardless, that non-members receive the water free, and that committee members often keep the fees for their own use. Association administration is impeded by members' ignorance of their rights or responsibilities, and the lack of co-ordination between *muang fai* operators and members. Water delivery is often inefficient or insufficient, the area covered by the association is too large to be overseen properly, and farmers are unwilling to work in formal organizations. Moreover, local politicking has destroyed the tradition of co-operation without compensation.[16]

Mechanization also aided farmers in breaking away from a centralized irrigation system. In the past, riparian farmers often built water-wheels to lift river water to their fields (Colour Plate 22; Plate 57). By the 1980s, many farmers were using diesel-powered pumps to draw water directly from the rivers and using long hoses to send it to distant fields. These methods eliminated reliance on gravity-fed canals and allowed farmers to bypass the water user associations.

57. The paddle water-wheel formerly used to move water from a canal to the fields is now propelled by a diesel motor or has been replaced by a pump and hose. (Courtesy National Archives)

It is clear that while the *muang fai* were inadequate for large-scale agriculture, neither have the government associations which replaced them succeeded in serving farm irrigation needs. Thailand's northern rivers are presently under considerable pressure to provide quantities of water from overstrained sources to serve a growing number of people. As there are no more stores of water to tap, the challenge is to make better use of existing supplies. Several possible approaches are discussed in succeeding chapters.

Major Lower-river Irrigation Schemes

Spurred by the development initiatives of King Chulalongkorn, Thai perceptions during the first half of the twentieth century changed from concordance with nature to domination of it. This period set the stage for the grand schemes undertaken at mid-century and the appearance of a new ethos espousing the exploitation of natural resources to bring the benefits of development to rural and urban Thais. It focused primarily on transforming raw materials into finished products for domestic consumption and for export. By the 1950s, forests would be felled for new, exportable wood products; minerals would be extracted and transmuted into industrial goods; and perishable agricultural products would be processed into goods with enhanced market values and shelf-life. This approach overturned the regimes that rivers had followed for centuries.

By the mid-twentieth century, dependence on agriculture as the foundation stone of the economy dictated that the river's annual peaks and ebbs be flattened to ensure a steady year-round supply of irrigation water. It reflected a belief that since nature is inconstant, man must impose his will upon it, exchanging sweat for certainty in creating reliable supplies of water. Steady growth required that the river behave itself and not erode its banks or inundate adjacent lands or towns.

A River Rampant

In part, the desire to tame the rivers arose from centuries of suffering its ravages. Floods occurred regularly every three or four years and were borne with equanimity by Thais ensconced in houses perched on stilts or floating on bamboo pontoons. Because the river seldom rose rapidly, villagers had ample time to move their goods and their animals to higher ground. As boats were the principal modes of transport, life continued virtually unchanged during the high-water period between mid-September and mid-November. But when heavy rains coincided with high tides, severe floods destroyed homes, livestock, crops, and people, and brought diseases. As the Thais moved further away from the Buddhist ideal of transitory existence and began building houses on solid foundations and transporting goods by road, the floods became a threat to prosperity. Rather than recognizing that permanence is seldom permanent and adjusting or even re-adopting the traditional approaches, Thais sought to stem the flow of the river with dams, and to contain it with embankments.

The first recorded devastating flood is detailed in the Dynastic Chronicles entry for 1785:

> In the same year, during the twelfth month [November], the water level was high everywhere, measuring up to eight *sok* (cubits), one *khup* (span), and ten *niu* ('fingers' or inches) [approximately 5.43 metres]. Many of the tender rice shoots intended for transplanting in the ricefields died. There was a shortage of food. The price of rice rose to one chang per kwian.... The king therefore commanded the Krom Na [Ministry of Agricultural Affairs] to dispense to the people a large amount of unhusked rice from the royal storage bins.[17]

During the floods of 1831, King Rama III erected a stone pillar at Ayutthaya (approximately 3.5 metres high) to gauge the water-level. While it is thought that flood levels prior to 1831—notably that of 1785—may have reached or topped that year's mark of 5.23 metres above sea-level, only that of 1942 has since approached it. In the two great flood years of 1879 and 1917 (other major floods occurred in 1871, 1882, and 1908), water rose to between 4.7 and 4.8 metres above sea-level. The flood of 1917 put Bangkok under water for three weeks and destroyed 464 000 hectares, or 21 per cent, of the rice crop in the Central Plains.[18]

The flood of the century, and the one remembered by Bangkok residents over sixty years of age, was that of 1942. Registering a maximum of 5.13 metres above sea-level on the Ayutthaya stone pillar, it was fuelled by heavy storms from China that caused the Chao Phya to rise 2.27 metres above normal in Bangkok, leaving Ratchadamnen Avenue under 1.80 metres of water. Bangkok inhabitants recall that food was scarce and that they survived by eating the fish that swam in and out of their windows. Men walked to work in waist-deep water, holding their pants over their heads, or were paddled in small sampans that had been pressed into service as taxis. There was no electricity, and the water from the taps was the same colour as that flowing past their doors. Snakes, rats, scorpions, and lizards invaded homes, and the Dusit Zoo's crocodiles escaped. On the positive side, burglaries and homicides dropped to practically nil and rats died by the thousands.[19] By the time the waters receded, 1 488 000 hectares, or 34 per cent of Thailand's cropland, had been destroyed. The floods were still fresh in the minds of the planners six years later when they contemplated the Greater Chao Phya Project to provide irrigation water and control flooding.

Water-borne Diseases

The warm tropical sun and the constant presence of stagnant water makes Thailand an incubator for disease. The chronicles and the tombstones of the foreign cemeteries bear epitaphs of lives severely curtailed by cholera, black-water fever (a form of malaria), breakbone fever (dengue), haemorrhagic fever, malaria, hepatitis, schistosomiasis, and other diseases which thrive in bogs and over-moistened land.

In past centuries, cholera, considered the greatest scourge, swept through Thailand in five-year cycles. The first recorded world cholera epidemic reached Thailand in 1820, progressing north from the mouth of the Chao Phya. After a number of deaths in Samut Prakan, residents fled to Bangkok and other cities, spreading the disease to the interior. In his 1828 *Journal*, Crawfurd recorded that the Phrakhlang (Minister of Finance, Trade, and Foreign Affairs) told him the epidemic was so devastating that

> it carried off two persons in ten, or a fifth part of the whole population.... The deaths ... were so frequent and sudden, that there was no time for the usual funeral rites, and the bodies were thrown in hundreds into the Menam; so that, according to their account, they presented the appearance of rafts of timber floating along the stream.[20]

The estimate of more than 35,000 corpses launched into the river does not include the thousands that would have been cremated. For a Buddhist people who abhor the disposal of the dead by any means but cremation and who fear the rampant ghosts of the uncremated, this method of disposal suggests the severity of the epidemic and the necessity for expediency. In later epidemics, the dead were carried through the Pratu Phi (Ghost Gate) in Bangkok's eastern wall, and exposed in the courtyard of Wat Saket for the vultures to devour. Even as late as World War II, a cholera epidemic killed 13,000 people. A flood control project was seen as an important means of halting the scourge's depredations.

New Perspectives

Like most South-East Asian rivers, the Chao Phya and its tributaries are monsoonal, dependent not on snow-melt from the mountains, but on rainfall. The rains fall in a very erratic pattern, causing the river to fluctuate between 'no flow' and overflow. It is commonly believed that Thailand enjoys a six-month monsoon season running from June through November. In truth, the season runs only three months, from September through November. Rains generally begin falling in the Chao Phya basin early in June, but aside from a few heavy showers, the volume is minimal. During July and August, rainfall tapers off, often to such an extent that crops wither in the fields. In drought years, crop failures generally occur during this period. After the 1950s, the new economic imperatives required that rivers be dammed, diked, and tamed to ensure an even flow and reliable volume year-round to provide electrical energy to power industry, water to irrigate crops, and to serve as a regulatory mechanism to reduce flooding.

Until the twentieth century, only two dams had been built: the thirteenth-century Saritphong Dam west of Sukhothai, and a dam on the Lopburi River near Phra Buddhabaht, constructed by King Prasat Thong in 1633 to serve a reservoir-based irrigation system for the central region. The first dam of modern times was the Rama VII Dam on the Pasak River at Tha Luang in Saraburi province. Thereafter, aside from the four small dams built on northern rivers, dam and canal construction had ceased.

The Greater Chao Phya Project

Twice in the nineteenth century, economic planners have targeted rice exports as the key to Thailand's prosperity. The first initiative, in the 1880s, spawned the Rangsit Project. After World War II, the Chainat Dam proposal was revived as the keystone of what was called the Greater Chao Phya Project. The impetus was a 1948 Food and Agriculture Organization study which concluded that Thailand's economic strength lay in exporting rice to alleviate world-wide food shortages caused by the war. Farmers were urged to plant higher-yielding rice strains and to double, or even triple, crop. The project would also supply water to serve nascent industrialization and growing urban populations. It was envisioned as a massive dam and canal complex extending to the outer edges of the lower basin (Map 13). Van der Heide's blueprints for a barrage at Chainat, rejected in 1903 and again in 1914, were dusted off and the RID was given responsibility for formulating the design and undertaking its construction. In 1950, Thailand secured an $18 million World Bank loan, and in 1952, commenced work on a twenty-five-year irrigation enhancement programme.

In stating its reasons for building the Chainat Dam and its trunk canals, the RID in 1957 reviewed the 117 years since the stone gauge at Ayutthaya had been erected. It noted that in 4 years unusually high floods had damaged crops. Conversely, in 60 years the lower valley had suffered drought, with 21 years of moderate drought, 35 years of severe drought, and 4 years of extreme drought, a total of 51 per cent of the entire period, 'by no means a small percentage'. In only 53 years had the water-levels been between 3.5 and 4.2 metres above sea-level so that farmers could depend on a good harvest.[21] It further noted that high-yielding 180-day rice required 1800 millimetres of water to produce a bountiful harvest (excluding an extra 800 millimetres to compensate for evaporation and canal leaks) and that normal rainfall was insufficient to provide it (Table 1). It concluded: 'It is clear therefore that without Irrigation, the Central Plain[s], the rice bowl of

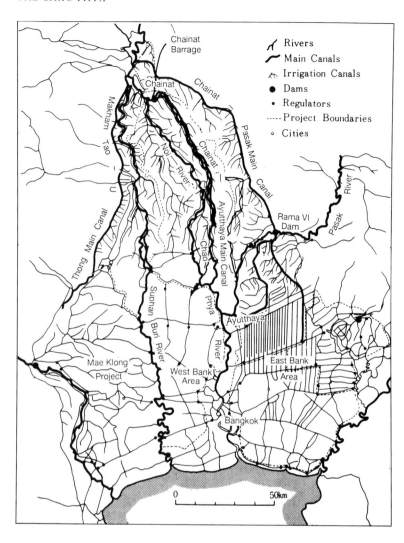

Map 13. The Greater Chao Phya Project.

Thailand cannot be made to produce the maximum yield it is capable of producing for the benefit of the rice-eating peoples of the world.'[22]

The RID enumerated the benefits of the Chainat Dam and attendant canal schemes as: supplying water for ploughing and sowing during May, protecting young plants from drowning in the first heavy monsoon rains and overflow from rain-swollen rivers in June, providing water to the growing crop in late July and early August when rainfalls and the river-levels drop, and overcoming the shortfalls or excesses already noted in its 117-year history of rainfall. Work began in 1953, and at its completion in 1957, the Greater Chao Phya Project was considered Asia's largest irrigation project. The dam, renamed the Chao Phya Dam, stretched 237.5 metres across the river and was fitted with 16 sluice gates, each 12.5 metres wide (Plate 58). Its maximum high-water discharge was 3300 cubic metres per second.

TAMING THE RIVER

TABLE 1
Rainfall Patterns (millimetres)

Region	Average Rainfall	
	Rainy Season	Whole Year
North (from west to east)	700–1000	800–1200
North-east (from west to east)	700–1000	800–1200
Central (from east to west)	1000–1100	1200–2000
South (east coast)	1500–3000	2000–4000
South (west coast)	2400–3000	3500–4000

Source: Royal Irrigation Department, *The Greater Chao Phya Project*, Ministry of Agriculture, 1957, p. 7.

58. The centrepiece of the Greater Chao Phya Project: the Chao Phya Dam at Chainat. It was the only major dam built with locks to facilitate inland navigation.

To extend irrigation facilities to the extremities of the basin, trunk canals were dug along the far edges of the valley. On the east, the Chainat–Pasak Canal was excavated from the Chao Phya Dam to the Rama VII Dam on the Pasak River (see Map 13). On the west, the Makham Tao–U-Thong Canal was excavated southwards from the dam to a point near U-Thong. The area between the eastern and western borders of the project would be served by the Tha Chin River, the Chao Phya Noi, and by a specially constructed trunk canal which ran from Chainat to Ayutthaya along the high left bank of the Chao Phya. The height of the latter

enabled it to feed the adjacent fields and distribution channels by gravity, even when the dry-season river-levels were low. During the rainy season, the five channels would also help draw flood waters away from the river, spreading them across the valley, thereby reducing the river's destructive force. This phase of the project became operational in 1962.

The third phase of the project was designed to control more precisely the levels of water delivered to the fields. This could be achieved only by constructing a network of capillaries to distribute water evenly throughout the entire basin. Thus, in 1962, the government promulgated the Ditches and Dikes Act, and assigned the RID the task of implementing it. Initially, farmers were asked to dig the canals but it was soon realized that with this approach, the project would take years to complete. Therefore, the RID undertook much of the work on its own. At its completion in 1969, the project provided gravity irrigation to 800 000 hectares of land via ditches dug at 400-metre intervals extending perpendicularly from the lateral channels. The ditches, each 1–4 kilometres long, were fitted with 'parshall flumes and vision boxes to ensure a uniform flow capacity of one litre per second per hectare'.[23] The project also controlled downstream flows during peak rainfall periods to eliminate or reduce flooding in Bangkok and other lower-river cities.

With the lower river effectively contained, the planners turned their attention to the northern tributaries. The *muang fai* and People's Irrigation organizations were sufficient for small-scale needs on minor rivers and streams, but for the degree of irrigation the government envisioned, only large dams would suffice. In 1952, Prime Minister Pibulsongkram had proposed a major dam on the Ping and once again the government turned to the RID to design and construct it. Rejecting a site in Yanhee Mountain, RID surveyors moved downstream to Khao Kaew (Crystal Mountain). Aided by hydraulic engineers from the US Bureau of Reclamation and the World Bank, dam construction began in 1957 utilizing a World Bank loan of $66 million, with the remaining one-third of the $100-million project secured from other sources. Both it and the subsequent Sirikit Dam were originally designed solely for water storage purposes, as opposed to irrigation schemes which would have involved the construction of lateral canals to carry the water to dry areas just below the dam. Only later were hydroelectric components added, a decision which would affect the manner in which water was allocated, and would contribute, in part, to the water shortages experienced in the basin since 1990.

On 17 May 1964, the arch-type Yanhee Dam (later renamed the Bhumibol Dam) was officially opened by His Majesty King Bhumibol (Colour Plate 23). At 154 metres tall, it ranked as the largest structure in South-East Asia, and the seventh largest dam in the world. Installation of additional turbines have raised its installed electricity generation capacity to 535 000 kilowatts, much of it des-

59. The Khiu Lom Dam on the Wang River above Lampang.

tined for Bangkok. The impounding efficiency of its 207-kilometre-long, 13.46 billion cubic metre reservoir ranks as one of the highest in the world in terms of net storage capacity.[24] A decision not to install navigation locks ended the Ping's history as the principal transport route between the North and the Central Plains.

The next major river to be dammed was the Nan. Just above Uttaradit, at the southern end of the Phasom gorge, construction began in 1965 on the Sirikit Dam. Completed in 1972, the 113.6-metre-high, 800.0-metre-long structure, Thailand's largest earthfill dam, can impound 9.51 billion cubic metres of water. Its hydroelectric generators have a total installed capacity of 375 000 kilowatts; a capacity enhancement programme would add 125 000 kilowatts by 1995. The dam irrigates 284 000 hectares of farmland in the Nan Valley during the rainy season, and supports dry-season agriculture for 48 000 hectares in the Nan River basin, and 400 000 hectares in the Chao Phya River basin.[25]

The Khiu Lom Dam on the Wang River just north of Lampang was completed in 1981 with a reservoir covering 112 square kilometres (Plate 59). Built as a multi-purpose dam, it also generates 400 kilowatts of hydroelectric power. The fourth tributary, the Yom River, is as yet undammed.

The Effects of Dam and Irrigation Projects

In the period between 1957 and 1982, four dams and more than 5000 kilometres of canals were constructed as part of the Greater Chao Phya Project. In addition, the ditches and dikes programme channelled irrigation water from the main arteries into individual fields. It is estimated that before the programme, transplanted rice accounted for only 22 per cent of the rice fields in the lower

Chao Phya Valley. This increased to 35 per cent in 1964, 42 per cent in 1967, 52 per cent in 1969, and 55 per cent in 1970.[26] By 1990, more than 70 per cent of the valley's farmers were engaged in irrigated agriculture.

The original projections of yield increases in the lower valley proved to be somewhat ambitious. It had been claimed that prior to 1950 harvests from fields relying solely on natural flooding had ranged between 625 and 1560 kilograms (average: 940 kilograms) per hectare and those from controlled flooding irrigation averaged 2060 kilograms per hectare. It was estimated that with the ditches and dikes programme, yields would rise to 2810 kilograms per hectare and, with the use of chemical fertilizers and stabilized water conditions, would increase to 3440–3750 kilograms per hectare, a virtual tripling. Harvests in parts of the delta did rise to 3750 kilograms per hectare, a figure which was presented as evidence of the programme's spectacular success.

Subsequent investigations, however, suggest that the 625–1560 kilograms per hectare yield estimates for naturally flooded rice may have been unrealistically low and that post-project gains in much of the basin were much lower than the projected figure of 3750 kilograms per hectare. In 1970, the average harvest reported by the RID for naturally flooded land was 1750 kilograms per hectare, and for irrigated rice fields in the Greater Chao Phya Project, 2940 kilograms per hectare, an increase of 68 rather than 300 per cent. In 1992, the total yield for rainfed land remained at 1750 kilograms per hectare while the total for irrigated areas had declined to 2705 kilograms per hectare (Colour Plate 24).[27]

Thus, while the project resulted in a sizeable productivity gain it was not of the magnitude forecast nor, perhaps, did it justify the enormous expenditure to build the system. Agricultural engineer Yoshihiro Kaida and others have suggested that the original figures were depressed for political reasons in order to spur relevant authorities to endorse the programme, a charge that was subsequently levelled by several observers to explain Thailand's present water shortages; that is, initial estimates of monsoonal runoff were inflated to rationalize construction of the major dams.[28] Furthermore, there has been considerable scepticism about the true role played by water in boosting yields. Leslie Small, in particular, suggests that the true potential of the Greater Chao Phya Project was never realized 'because the existing system does not have the capacity to support a true revolution in production techniques'.[29] The principal value of the improved transplanted rice varieties was not a matter of higher yields but of security; if rainfall is insufficient during the crucial month of July, the farmer only has to replant the seed-bed, not the entire crop. What is certain is that the two decades between 1960 and 1982 saw a dramatic rise in the amount of newly irrigated land, from 1.30 million hectares in 1957 to 3.33 million hectares in 1982. Since the mid-1980s, the RID has brought a further 1.20 million hectares under irrigated cultivation.

In the flurry of attention over the yield increases, the question not raised concerned the dramatic increase in demand for water, a legitimate consideration in the light of the water shortages the lower basin is presently experiencing. The shortages owe their origins to the stunning success of the Greater Chao Phya Project, which achieved its objectives on a scale unprecedented in Thai history. The project had been initiated to support propagation of high-yielding varieties of wet-season rice which required higher water inputs than local rice varieties, and higher than rainfall alone could provide. But with the promise of new prosperity, farmers switched to the new varieties in such numbers that the demand for water began to climb precipitously. By the 1980s, the system was no longer able to provide irrigation water in the amounts necessary to supplement the rainfalls, and conflicts between farming zones began to arise. Moreover, the distribution of water throughout the system was uneven with the fields nearest the canals receiving most of the water, leaving distant areas without irrigation.

In addition to the increase in high-yield primary-season rice plantation, expansion by farmers of second-crop rice propagation exceeded targets. Technical limitations in water distribution, however, meant that even under optimum conditions, only one-quarter of the lower basin could be brought under dry-season rice cultivation. Furthermore, additional costs for fertilizer and other inputs made the propagation of higher-yielding dry-season rice varieties economically unfeasible for many farmers.

The project also proposed to provide water to expand dry-season field crop cultivation, primarily along the northern portion of the basin. A ten-year plan begun in 1965 would raise the dry-season cultivation area by 16 000 hectares per year during the initial five years, 28 000 hectares in 1971, and 24 000 hectares per year for the final four years. New crops, including upland beans and nuts, vegetables, fruits, and sugar-cane, would be introduced. The project failed because of several factors, many of them human. Trunk and lateral channels had been insufficiently developed in part because government departments were not co-ordinated, land consolidation necessary for success was not carried out, soils were unsuitable except along embankments, variable rice harvest times hampered vegetable planting schedules, farmers lacked suitable machinery and implements, vermin and pests became a problem, and credit and marketing opportunities were limited.

The scheme's effectiveness has also been hamstrung because, as in the northern water user associations, farmers were not included in the decision-making process. Central Plains farmers were assigned to user groups but they failed to function as unified associations, in part because the initiative came from a central authority. By failing to educate the farmers on the benefits of co-ordinating their efforts, the government left itself vulnerable when conflicts over water allocation occurred, as they would with increasing frequency in the water shortages of the 1990s.

The Lower Portions of the Four Tributaries

While northern valley water needs have been served by RID and *muang fai* projects, and the lower basin has been supported by the Greater Chao Phya Project, the portions of the Chao Phya system between the Sirikit and Bhumibol Dams and the Chao Phya Dam required special measures. Since the mid-1980s, the RID has been excavating lateral canals to carry water from the river into adjacent fields. Equipped with gates and relying on gravity feed, these channels are effective only during the rainy season when the river-levels rise (Plates 60–61). As such, they also serve as dispersion channels to alleviate flooding in downriver areas. During the dry season, farmers along the lower tributaries plant crops in the silty soil enriched by the previous monsoon season's deposits of upstream topsoil.

Numerous small barrages have also been built on the four tributaries to raise reservoir levels so that irrigation water can be sent by gravity flow down concrete channels to fields dozens of kilometres away. An improvement on the concrete dam is the *khuan yang* (rubber dam) in which a rubber bladder is anchored to either bank and stretched across a river. Hydraulic machinery fills the bladder, raising its height to block the passage of water. This adjustable dam can be raised or lowered up to 2 metres as needs and seasons dictate.

In regions between barrages, where the river-level is several

60. One of dozens of small irrigation dams with gates; this one is on the Pasak River.

TAMING THE RIVER

metres below the banks or where soil composition does not support a permanent dam, the RID has excavated canals at right angles to the rivers. High-tension electrical cables run from power-stations to large floating pump houses (Plate 62) anchored in the river below the riverside-end of a canal. The electric pumps draw water from the river through pipes more than 30 centimetres in diameter, raising it to the level of the canal and sending it to distant fields. In 1990/1, 159 pumps, including 53 in Uttaradit province in the Central Plains, were irrigating 49 440 hectares, double the total land area irrigated by pumps in 1985/6. In the North, 46 pumps were operating in 1990/1 (25 in Lampang province alone), irrigating 11 040 hectares, triple the 1985/6 levels.[30]

While the construction of the Chao Phya, Bhumibol, and Sirikit Dams sharply reduced the threat of widespread flooding, they were unable to cope with the heavy rains that resulted in severe flooding in 1975, 1983, 1986, and 1990. The flood of 1983 put several Bangkok districts, notably those in the Hua Mak area, under water for up to six months and caused destruction calculated at B6,600 million ($253 million).[31]

Neither have the dams been sufficient barriers to moderate flooding which has increased since 1964, aided in large part by the gradual sinking of the city itself. As a result, annual flooding is common in the lower valley, exacerbating the city's traffic problems, interrupting commerce and communications, and damaging property. On 29 September 1988, a rise from the normal 3.0 metres to

61. Watergates control floods and provide irrigation water on the lower portions of the Chao Phya tributaries.

62. A floating pump house on the Nan.

6.2 metres on a Chao Phya distributary, the Tha Chin, was not contained by control canals at Sam Chuk and Don Chedi. As a result, the waters inundated 58 720 hectares of farmland, devastating the rice crop. Even the northern areas protected by barrages did not escape severe flooding. In Chiang Mai, Typhoons Betty and Cary in August 1987 created the city's worst floods in twenty years with the Ping River rising to 4.55 metres, 26 centimetres higher than the levels of the 1965 flood, and putting main streets under 0.5–1.5 metres of water. The same storms raised waters in Khiu Lom Dam, threatening residents in the Wang River. In both instances, forests which would have absorbed some of the runoff had been felled.

Since the 1960s, Thailand has experienced a dramatic rise in the incidence of water-borne diseases. Stagnant water left after the 1983 flood in Hua Mak served to spread a number of ailments including hepatitis. World-wide, irrigation projects have been blamed in the epidemiology of more than thirty water-related diseases which have proliferated in farm areas.[32] Twenty-five of those infections occur in Thailand, including cholera, malaria, typhoid, schistosomiasis, dysentery, and meliodosis.[33]

Thus, despite minor improvements in yields brought about by massive irrigation and dam-building schemes, it seems clear that these projects have created as many problems as they have solved. While it is admirable that the government achieved its goal of making Thailand the world's premier exporter of rice, it is that very success that has created water shortages of a magnitude that threaten Thailand's future prosperity. Rising water demands by all sectors of the economy are placing enormous pressure upon the Chao Phya River system, and threatening its very viability. Resolving this dilemma requires major initiatives, not to seek new sources of water, but to improve management of available supplies. That task requires first examining a number of constraints on implementing successful, comprehensive water management programmes.

1. Chamnongsri Lamsam Rutnin, 'Nature in the Service of Literature', in *Culture and Environment in Thailand: A Symposium Sponsored by the Siam Society*, Bangkok: Siam Society, 1989, p. 245.
2. Shigeharu Tanabe, 'Land Reclamation in the Chao Phraya Delta', in Yoneo Ishii (ed.), *Thailand: A Rice-growing Society*, trans. Peter Hawkes and Stephanie Hawkes, Honolulu: University Press of Hawaii, 1978, p. 57.
3. Piyanart Bunnag, Duangporn Nopkhun, and Suwattana Thadaniti, *Khlong Nai Krungthep*, Bangkok: Chulalongkorn University, 1982, Supplement V.
4. Ibid.
5. Ian Brown, *The Élite and the Economy in Siam c.1890–1920*, Singapore: Oxford University Press, 1988, pp. 11–12.
6. James C. Ingram, *Economic Change in Thailand 1850–1970*, Stanford: Stanford University Press, 1971, p. 85.
7. Brown, *The Élite and the Economy in Siam c.1890–1920*, pp. 19–20.
8. Ibid., p. 30.

9. Ingram, *Economic Change in Thailand 1850–1970*, pp. 199–200.
10. Vanpen Surarerks, *Historical Development and Management of Irrigation Systems in Northern Thailand*, Chiang Mai: Department of Geography, Chiang Mai University, 1986, p. 208.
11. Ibid., p. 101.
12. Ibid., p. 118.
13. Jack M. Potter, *Thai Peasant Social Structure*, Chicago: University of Chicago Press, 1976, p. 100.
14. Vanpen, *Historical Development and Management of Irrigation Systems in Northern Thailand*, p. 237.
15. Richard P. Lando, 'The Spirits Aren't So Powerful Any More: Spirit Belief and Irrigation Organization in North Thailand', *Journal of the Siam Society*, Vol. 71, Pt. 1, 1983, p. 140.
16. Vanpen, *Historical Development and Management of Irrigation Systems in Northern Thailand*, pp. 422–3.
17. Thadeus Flood and Chadin Flood (eds., trans.), *The Dynastic Chronicles Bangkok Era, the First Reign: Chaophraya Thiphakorawong Edition*, Tokyo: Center for East Asian Cultural Studies, 1978, Vol. 1, p. 86.
18. Royal Irrigation Department (RID), *The Greater Chao Phya Project*, Ministry of Agriculture, 1957, pp. 8–9.
19. Rungsi Prachonpachanuk, 'Bangkok under Water', *Bangkok Standard*, 31 August 1969, p. 9.
20. John Crawfurd, *Journal of an Embassy to the Courts of Siam and Cochin China*, London, 1828; reprinted Kuala Lumpur and Singapore: Oxford University Press, 1967 and 1987, p. 455.
21. RID, *The Greater Chao Phya Project*, pp. 8–9.
22. Ibid., p. 10.
23. Yoshihiro Kaida, 'Irrigation and Drainage, Present and Future', in Yoneo Ishii (ed.), *Thailand: A Rice-growing Society*, trans. Peter Hawkes and Stephanie Hawkes, Honolulu: University Press of Hawaii, 1978, pp. 220–1.
24. Electricity Generating Authority of Thailand (EGAT), *Bhumibol Dam and Hydropower Plant*, Bangkok, 1993.
26. Yoshihiro, 'Irrigation and Drainage', p. 226.
27. Agriculture Ministry statistics for 1992.
28. Yoshihiro, 'Irrigation and Drainage', p. 227.
29. Leslie Small, 'An Economic Evaluation of Water Control in the Northern Region of the Greater Chao Phya Project of Thailand', Research paper, South East Asia Development Advisory Group, Washington, DC, 1971.
30. Center for Agricultural Statistics, *Agricultural Statistics of Thailand, Crop Year 1991/92*, Office of Agricultural Economics, Ministry of Agriculture and Cooperatives, Bangkok, 1992, pp. 208–9.
31. Anat Arbhabhirama, Dhira Phantumvanit, John Elkington, and Phaitoon Ingkasuwan, *Thailand: Natural Resources Profile*, Singapore: Oxford University Press, 1988, p. 137.
32. Jose Olivares, 'Health Impacts of Irrigation Projects', in G. Le Moigne, S. Barghouti, and H. Plusquellec (eds.), *Dam Safety and the Environment*, World Bank Technical Paper No. 115, Washington, DC: World Bank, 1990, pp. 149–64.
33. Wellcome Unit, Epidemiology Centre, Mahidol University, Bangkok, 1993.

7 River under Siege

The Central Area is aptly called by the Thai, in their idiom, the storehouse of rice and water.[1]

The Critical Shortfall in Water Supply

THAI scholar Phya Anuman Rajadhon's assertion reflects the traditional acceptance by Thais of the liquid largess which nature has bestowed upon them. Had prognosticators of old peered into the 1990s, it is likely that they would have seen the Chao Phya and its tributaries brimming with water and prosperous farmers labouring in thick stands of lush green rice. It has become apparent, however, that the storehouse's bounty can no longer be taken for granted. Thailand is beset by serious water shortages that become more severe with each passing year. The river system is imperilled by the volumes of water being extracted from it and the pollutants being poured into it. Were Thailand at this moment to settle into a zero growth economy, it might be able to remedy the shortfalls with relative ease. But the dynamic economic growth of the 1980s has continued unabated into the 1990s and shows no signs of flagging, further exacerbating the rivers' woes.

The shortfalls are of such magnitude that in 1994, the Thai government was advising farmers to expect their irrigation supplies to dry up (Plate 63), and warned Bangkok residents that they

63. A Central Plains canal during the hot season.

64. The Ping and Nan Rivers at Nakhon Sawan in January 1988.

faced water rationing, a situation unprecedented in Thai history. Thailand had suffered water shortages before but never on the scale of those which began appearing after 1988 (Plates 64–65).

In April 1989, the government asked lower basin farmers to refrain from planting a dry-season crop because the reservoirs could not supply their fields and still have sufficient water for the wet-season rice crop the following July. Instead, farmers planted 160 000 hectares more land than available water supplies permitted.

65. The Ping and the Nan at Nakhon Sawan in January 1994.

While it did release some water, the Royal Irrigation Department (RID), in effect, consigned the second crops to their doom. Even then, there was not enough water for the subsequent primary rice crop.

Reservoir levels began falling and continued to fall over the next four years. In September 1993, during the peak rainfall month, the RID reported that water-levels were so low at the Bhumibol and Sirikit Dams that instead of the 30–40 million cubic metres of water per day normally released at that time of year, it was releasing only 17 million cubic metres per day. The Electricity Generating Authority of Thailand (EGAT) shut down turbines at both dams and compensated for hydroelectric shortfalls by switching to its coal and natural gas electricity generating plants. On 1 December 1993, the RID banned the planting of second crops until the following May, thereby contravening the very purpose for which the Greater Chao Phya Project had been undertaken. It claimed that it took the step to conserve water for boat transport, flush river pollution out to sea, and to counter sea-water intrusion into the lower river. It warned that a critical point was being reached in which the government would have to choose between serving farm or urban needs because it could no longer adequately provide for both.

Reasons for the Shortages

In the subsequent analysis, a number of reasons were postulated for the dilemma. Some government agencies contended that the water shortages are cyclical, caused by droughts that occur every 3–6 years, and that there was no undue cause for alarm. The years 1967–8, 1972, 1977, 1979, 1986–7, and 1990 had been drought years and water had been rationed. But while temporary shortfalls formerly occurred in the period immediately following a drought year, present shortages appear to be chronic and of longer duration than in the past. A look at average annual water-levels over a fifteen-year period for the Bhumibol and Sirikit reservoirs for the peak month of November and low month of June reveals the extent of the problem (Table 2).

The low reservoir levels led to urban shortages and sparked fierce conflicts between regulatory agencies and farmers and between competing users. In November 1992, Bangkok residents in some areas had paid up to B400 ($18) for a 1000-litre-tank of water, and two groups of farmers in Chainat province had engaged in a 'free-for-all' over control of water in their district. RID officials in Suphan Buri and Sing Buri were threatened when they tried to conserve river water for general use. In Nan and Phrae, 'influential' tobacco farmers had hired security guards to protect their water sources.[2]

In January 1993, the Director of the RID offered four reasons for the water shortages: lack of rain in the catchment areas of the

TABLE 2
Declining Reservoir Levels at Key Dams

	Bhumibol Dam		Sirikit Dam	
Year	Reservoir Level (metres above sea-level)	Storage (million m³)	Reservoir Level (metres above sea-level)	Storage (million m³)
High Level (November)				
1978	250.81	10700	159.75	7782
1982	248.47	10056	156.57	8206
1984	244.60	9077	157.00	8260
1987	239.65	7923	140.72	4780
1988	250.40	10611	149.99	6647
1989	241.84	8414	149.07	6448
1990	231.42	6403	143.57	5319
1991	234.90	6979	140.53	4746
1992	235.75	7135	141.51	4926
Low Level (June)				
1979	235.95	7191	140.27	4697
1982	238.81	7778	142.39	5073
1985	233.73	6774	141.38	4902
1988	230.25	6234	136.28	4012
1989	233.95	6812	142.39	5092
1990	223.46	5190	137.43	4202
1991	216.65	4254	132.07	3375
1992	213.84	3901	129.20	2997
1993	218.58	4507	132.26	3402

Source: Electricity Generating Authority of Thailand.

two dams, increased and wasteful use of water in the lower Chao Phya basin, increased population and consumption upstream from the dams reducing flows into the reservoirs, and a general increase in cultivation areas. He noted that only 20 per cent of the main river flow was captured by dams with 80 per cent flowing unused into the sea. His assessment failed, however, to include a number of other important factors.

The low reservoir levels have been attributed to substantial declines in rainfall in recent years, yet Meteorological Department figures do not support this contention. Five-year averages reveal that rainfall decreases have been relatively minor, and in some areas, rainfall has been higher than in the past (Table 3). The moderate decline merely reflects the two normal cycles of rainfall, the first occurring every 60 years and the second every 11–12 years. Thailand is now in a trough but by 1995 annual rainfall averages should begin to rise again. Thus, attributing the declining reservoir levels to dwindling rainfall is insupportable. The decreasing volumes are caused by the increased use of water by farmers living upstream from the dams. That contention is evident

TABLE 3
Annual Rainfall Records (millimetres)

Year	Phitsanulok	Tak	Nakhon Sawan	Bangkok	Chiang Mai	Nan	
1978	919.1	881.0	1208.5	1236.4	1350.8	1237.1	
1979	938.9	877.7	678.3	1133.4	967.8	957.2	
1980	1768.9	1225.0	1222.5	1471.0	1224.0	1547.6	
1981	1308.6	1257.8	1451.6	1592.7	1218.8	1235.0	
1982	1179.7	838.6	998.0	1829.6	836.8	1132.8	
1983	1627.1	1368.5	1199.8	2129.5	1026.9	1391.9	
1984	1011.4	972.4	886.0	1398.0	783.8	1332.9	
1985	1649.1	1111.7	1197.9	1368.7	1225.9	1191.5	
1986	1236.8	1094.4	897.3	1807.5	984.2	985.9	
1987	1147.5	978.5	1123.0	1370.3	1144.5	928.5	
1988	1324.8	1226.4	1618.3	2097.3	1412.4	1373.5	
1989	1398.9	1092.4	840.1	1496.4	1190.4	922.2	
1990	1045.9	946.8	1054.5	1362.9	1161.0	1269.8	
1991	958.4	909.7	608.5	1358.5	1006.8	962.6	
1992	1129.7	968.1	983.9	1435.1	1039.3	1105.5	
Five-year Averages							Average
1978–82	1223.0	1016.0	1111.8	1452.6	1119.6	1221.9	1190.1
1983–7	1334.4	1105.1	1060.8	1614.8	1033.1	1166.1	1219.0
1988–92	1171.5	1028.7	1021.1	1550.0	1162.0	1126.7	1176.7
Volume Changes over Previous Five-year Period							Average
1983–7	+111.4	+89.1	−51.0	+162.2	−86.5	−55.8	28.9
1988–93	−162.9	−76.4	−39.7	−64.8	+129.0	−39.4	−42.6
1988–93 cf. 1978–82	−51.5	+12.7	−90.7	+97.4	+42.4	−95.2	−13.5

Source: Meteorological Department, Royal Thai Government.

in the drastically reduced inflows into the two main dam reservoirs, as Table 4 suggests. That is further supported by RID statistics showing that the average annual volume of water flowing along the Chao Phya River measured at Nakhon Sawan over the twenty years between 1972 and 1992 was 22 190 million cubic metres. During 1982–92, that had declined to 18 700 million cubic metres. For the five years between 1987 and 1992, the annual average was 16 700 million cubic metres suggesting that by the end of a twenty-year period, 5500 million cubic metres never entered the lower river system but were channelled into fields by farmers living north of Nakhon Sawan.

Inflows are substantially reduced by evaporation which, although not as severe as in arid climates, amounts to 1 cubic metre of 'free water' surface area per year. For example, of the 370 million cubic metres flowing into the Bhumibol and Sirikit reservoirs between January and June 1993, the RID expected that 200 million cubic metres, or 54 per cent, would evaporate. A portion of the evaporated water returns to the system as rain that showers distant

TABLE 4
Inflows into Dam Reservoirs (million cubic metres)

Fiscal Year[a]	Bhumibol Dam	Sirikit Dam
1982	5907	5082
1988	5441	4585
1990	4456	4132
1991	4270	3449
1992	3758	2926
1993	3415	3467

Source: EGAT Reservoir Records, Systems Operations Department.
[a]October–September.

hillsides but much of it falls on countries north of Thailand. Unfortunately, there is no viable technology to slow the evaporation rate.

Low reservoir levels stem in part from initial miscalculations of the rivers' hydrology. In the early 1960s, the RID estimated that an annual runoff of 1.8 billion cubic metres of water would flow into the Bhumibol and Sirikit reservoirs. After the dams' completion, however, the inflows amounted to only 1.2 billion cubic metres per year. Some geologists suggest that two fault lines which run across the Bhumibol reservoir might account for leakage but hydrologists contend that water systems are closed, meaning that any water disappearing underground at one point has to resurface at another point. There is no evidence that large quantities of water have welled up elsewhere, so the theory of leakage through a crack in the reservoir floor must be discounted. Several observers have argued that the miscalculations were deliberate in order to ensure that the dams would be built, a manœuvre not unknown in preparing decision-makers to approve and finance river engineering programmes.

These miscalculations hindered officials in meeting irrigation demands in another vital agricultural area. Although water provision for dry-season cropping was a key objective of the Greater Chao Phya Project, by 1978, only one-quarter of the land targeted had been brought under cultivation, for reasons already outlined. Of greater concern, though, was the fact that there were insufficient supplies to irrigate even that one-quarter. Several explanations have been forwarded for the discrepancy, notably the lack of reliable data at the planning stage prior to the project's initiation. This lacuna was exacerbated by planners' failure to make realistic projections of future water needs and uses, or to deal with the independent nature of the average Thai farmer unaccustomed to acting in concert with other farmers or in accordance with a national water management plan. Lack of communication between government agencies and farmers, insufficient co-ordination among regulatory agencies, and a failure to establish who would receive the water created a situation wherein unregulated water use by one sector deprived intended users in other sectors.

Thus, although the Sirikit Dam had been constructed in part to provide northern basin farmers with water for dry-season crops, the availability of water encouraged lower basin farmers to plant high-yielding dry-season rice and in such quantities as to stretch limited water supplies. RID figures reveal that land planted with dry-season rice rose from 40 per cent of cultivatable land in 1975 to 60 per cent in 1979, and more than 70 per cent by the late 1980s.[3] It created a situation whereby in the early 1990s EGAT was simply obliged to cut off water supplies to overreaching farmers.

This situation reflects a serious shortcoming in approaches to water management. In its development projects, government agencies tend to emphasize physical structures and to give insufficient consideration to human factors. Relevant agencies compete with each other rather than co-ordinating their efforts. Policy is generally formulated without consultation with, or adequate explanations to, end-users. Little effort is exerted to encourage farmers to use water efficiently by teaching them proper conservation techniques, or to convey the importance of canal maintenance in reducing leakage and ensuring steady flows. The farmer has been led to believe that responsibility for maintaining the system lies with the RID, but the agency lacks the resources to carry out such duties effectively. As a consequence, farmers waste water, compete rather than co-operate, and have an unclear idea of objectives or the value of a national water management programme that could ensure them adequate supplies. The RID's response to shortages has been to press for the construction of more dams.

Complicating Factors

Before examining the wisdom or feasibility of expanding the water supplies, it is important to consider a number of other factors which imperil the Chao Phya and impinge on its ability to serve the basin through which it flows. In the mid-1990s, the river system looks little like it did in past centuries. In addition to the obvious physical alterations, like *khlong lat* short cuts and dams, there are numerous small changes which damage the river and confound remedial measures. In many upstream areas, notably on the Wang and the Yom, barrages tap so much water for farm irrigation that even during the monsoon season, the river below the dams is a bare trickle and, in some instances, is little more than a moist bed. Downriver portions are polluted by farm chemicals, garbage, dead animals, and human waste.

Chapter 5 explored the various ways in which Thais perceive rivers, and the beliefs that dictate how they utilize them. It is clear that the river is regarded, not as a mystical entity, but simply as a ditch to convey water for farm, urban, and industrial purposes. Worse yet, it is regarded as a sewer to the sea into which waste can be flushed and forgotten. Ultimately, these acts reflect a public perception that the river is an anachronism, no longer important to Thailand's destiny. Yet, as is becoming increasingly evident, the river is vitally important, perhaps more so than in the past.

In simpler times, river abuses were on such a small scale that their impacts were negligible. Pollutants and waste matter, composed primarily of biodegradable materials, were poured into it and were carried away by the current; village middens on the river-banks were scoured clean by the annual floods. A river could be dredged, dammed, re-channelled, and its watersheds felled with little concern for the consequences to downstream areas. But rising population and the over-exploitation of resources are taking their toll, creating such a multitude of environmental abuses that planners are now openly suggesting that if nothing is done to improve conditions, the Chao Phya system may cease to exist in as little as two decades. In September 1993, the Food and Agriculture Organization warned that Thailand might not be able to feed its population in the twenty-first century. It noted that fish production had levelled out at 97 million tons, grain output had fallen by 8 per cent from its 1984 peak, and meat production had declined. It cited widespread damage to biodiversity as the principal factor limiting the country's ability to increase production or even to maintain present levels. An examination of the myriad problems confronting the country reveals that such views are not alarmist; indeed, they show a clear-sighted perception of the future.

Contrary to popular opinion that river degradation is perpetrated solely by Bangkok residents and factories, rampant abuses in myriad forms take place as high up the river as humans dwell; in some instances, such as the Wang River, to the very point where the headwaters emerge from the earth. Given the state of the upper river today, it seems difficult to credit a 1688 comment by

66. Detergent in the Ping River at Chiang Mai.

TABLE 5
BOD Levels in the Chao Phya River (milligrams/litre)

Course	Standard	1987	1988	1989
Upper	1.5	1.6	1.7	1.0
Middle	2.0	1.8	1.8	2.4
Lower	4.0	4.0	3.8	2.8

Source: National Environment Board, Thailand.

Nicolas Gervaise on the purity of the Chao Phya River as it passed Ayutthaya: 'The water of the river [Chao Phya] is extremely clear, light and excellent to drink.'[4]

Today, it is not unusual to travel the upper rivers and find the water frothing with washing detergents (Plate 66), or trees along the banks festooned with plastic bags, old clothing, and discarded plastic motor-oil bottles. Towns add garbage and human waste; farms add chemicals. As noted in Table 5, Biological Oxygen Demand (BOD) counts are abnormally high along the entire length of the river.

Pesticides

One of the prime components of the 1960s 'Green Revolution' in agriculture was the use of chemical pesticides and herbicides. While these eradicated harmful insects and weeds, errors in application—stemming from improper explanations to farmers—have had a twofold impact: contamination of soils, and pollution of rivers from chemicals washed into them by erosion and rainstorms. It is estimated that 25 per cent of agricultural water is returned to the system, carrying with it pesticides and fertilizers. The streams and rivers are also contaminated by farmers who wash their spray tanks in them.

A 1988 study by the Provincial Waterworks Authority found levels of dieldrin (the prime toxic ingredient in pesticides) in river water higher than the 0.05 parts per million standard set by the US Environmental Protection Agency. These chemicals poison fish and pollute downstream water used by riverine dwellers for cooking, bathing, and teeth brushing. The Watershed Hydrology Department claimed that in 1985, 24 people died and 2,500 fell ill from farm chemicals, and that each Thai inadvertently consumes at least 1 pound (0.45 kilogram) of pesticides and other poisonous chemicals each year.

Deforestation

The loss of forest cover and destruction of watersheds have adversely affected river regimens, reducing the ability of the soil to hold rain-water and release it gradually over time. It has also contributed to the loss of topsoil through erosion. Despite a logging ban imposed in 1989, trees are still being cut. Surveys in 1961, revealed that 53.33 per cent of Thailand's total land area was forested; in 1987, that figure had dropped to 28.03 per cent. Other surveyors have speculated that, discounting fruit orchards and coconut plantations, the true figure is closer to 20 per cent while the most pessimistic claim that it may have dropped as low as 12 per cent.

Chiang Mai province, the watershed for the Ping River and its half-dozen tributaries, is typical of most northern provinces. Royal Forestry Department surveys in 1992 revealed that Chiang Mai's forests had shrunk by 10 per cent since 1981 when 1.7 million hectares of trees covered the province. By 1990, land under irrigation had increased by 300 000 hectares. It could be argued that, ironically, the RID has contributed to deforestation. By providing farmers with more water to irrigate crops, it inadvertently encouraged them to clear more land to increase their incomes (Plate 67).

Blame for deforestation has been placed on hilltribes clearing land for swidden or slash-and-burn agriculture, and on villagers or migrants seeking new land to till, but the poachers range over a wide variety of income groups. It has also been documented that forests are illegally cut by gangs hired by influential sawmill owners operating with the tacit approval of, and often in partnership with, provincial police.[5] At the other end of the spectrum,

67. Hills above the Ping denuded to plant cabbages.

trees are damaged and ultimately destroyed by villagers scraping the bark from *kho* trees to obtain a fibre that prolongs the flavour of betel-nut.[6] Lower in scale than the loss of forests, but equally important is the destruction of bamboo to serve tourism. On the Mae Taeng River alone, an estimated two million one-year-old bamboo trees are cut each year to build tourist rafts.

Considerable damage is also done by fires. Northern skies in January are palled with smoke and ash as hillside fires ignited by farmers consume fallen leaves and duff. The trees, still moist from monsoon rains, are left unscathed. The fires ensure that the dry ground cover does not become tinder to fuel a forest fire during the subsequent hot season (Colour Plate 25). In the Central Plains, harvest stubble is set ablaze as the only non-chemical means of controlling a fungus which infests rice fields and can damage the following year's crop. In both instances, the fires bake the clayish earth, reducing soil fertility and inhibiting its ability to absorb water.

Mimosa

Water weeds and reeds have always flourished along river-banks but in recent decades a more serious problem has appeared. The thorny *Mimosa pigra* L. (*Mimosaceae*) or Giant Mimosa (*ton mayarap yak* in Thai) grows in dry areas but thrives in water margins, especially in areas exposed to bright sunlight (Colour Plate 26). Known to reach heights of 7 metres, it can grow in water up to 1 metre deep, meaning it can extend a considerable distance into most waterways. It also chokes river-banks and invades fields, its extensive root system resisting all efforts to extirpate it.

It has become so pervasive that farmers who formerly cultivated the river-banks after the monsoon waters had receded have ceased all efforts to clear it and as a result, have lost a considerable quantity of valuable crop land, especially along the tributaries. It has so infested the upper Wang River that it forms a wall, completely blocking passage along a 40-kilometre section between Wang Nua and Chae Hom. A similar situation exists along the upper Yom River above Amphoe Pong. Farmers claim that the infestation began in the early 1980s but the convening of a special conference in Chiang Mai in 1983 suggests it already constituted a major problem at that point. The plant's history provides an intriguing study of conflicting interests that infect policy formulation along major portions of the river.

In 1947, the mimosa was purposely introduced into northern Thailand from Indonesia as a green manure and cover crop in tobacco plantations. It was later planted to help control ditch-bank erosion around Chiang Mai. After 1975, it quickly spread upriver into Chiang Rai and Chiangsaen, its seeds carried in road construction sand dredged and transported from the lower river. After infesting Lamphun, Lampang, Tak, and Kamphaeng Phet, it moved downriver, reaching Bangkok by 1980.[7] By 1983, it covered 18 800 hectares along the Ping River. Assuming that mimosa extends 20 metres back from either bank, that means it infests 4700 kilometres of stream and river-banks, a not inconsiderable area.

In addition, 800 hectares of the Khiu Lom Reservoir and 810 hectares of irrigation systems were invaded. In the 1950s, the RID planted mimosa to control bank erosion on dam reservoirs, and the plant subsequently spread to cover vast areas. Ironically, although it protects against bank erosion, it also promotes sediment build-up which reduces the reservoir's useful life. For example, it is estimated that its presence will reduce by one-quarter the life of the Khiu Lom Reservoir, originally projected for 100 years.[8]

The weed is particularly noxious along canals where it blocks the passage of sediment and, ultimately, of irrigation water. It has been suggested that it has the potential to virtually eliminate agriculture relying on irrigation. Growing primarily on the inside bends of rivers, its roots and stems increase sediment deposition by an estimated 75 per cent, causing the more rapid erosion of the opposite (outside) bend.

Mimosa has several uses—as firewood, beanpoles, and compost—but its price is so low that cutting it commercially is economically unfeasible. Efforts to eradicate it have been fruitless. It is extremely prolific with a single plant producing between 42,000 and 98,000 seeds a year after 6–8 months. The seeds are drought resistant and can remain submerged for at least 30 days before dying; seed pods submerged for more than 2.5 years will still germinate. Thus, raising the water-level in reservoirs will not kill it.[9] This leaves two control methods: mechanical (cutting, digging, or burning) and chemical (herbicides, especially aerial spraying just before the rainy season). The herbicides cause the foliage and thorns to drop off, making it easier for farmers to harvest the stems, but neither method is viable for large-scale eradication programmes.

Two new techniques to control the mimosa are presently being field tested. The National Biological Control Research Centre at Bangkok's Kasetsart University is using two species of seed-eating beetles—*Acanthoscelides puniceus* and *Acanthoscelides quadridentatus* (Coleoptera: Bruchidae), which were imported into Thailand from South America, after extensive testing in Australia. They curtail the mimosa's proliferation by eating its seeds before they mature. Since August 1983, beetles released at selected northern sites have inhibited seed production by 15–35 per cent; in some areas of Chiang Mai, they have reduced seed maturation to zero.[10] Because the beetles only diminish seed production, they can control the mimosa's spread but cannot eliminate existing plants. Their benefit is that they are host-specific so will not damage other plants. The Centre is also experimenting with a species of Diabole fungus which attacks the plant, causing die back. Initially certified in South America, and presently being tested in Australia, it may be introduced in Thailand in 1994. Unfortunately, it works only in dry areas, necessitating a search for a species which will attack river-bank and moist-area mimosa.

There is, however, considerable opposition to mimosa eradication programmes. While in the 1980s both the RID and EGAT came to recognize the magnitude of the problem and have carried out control programmes using chemicals, its cultivation is actively promoted by bee-keepers because it blossoms year-round. In addition, there is fierce opposition to the enactment of laws similar to those in other countries which control noxious weeds and require that landowners remove them from their lots. Developers and others who have purchased land as investments oppose any measure which would require expenditure to control the proliferation of mimosa.

Thus, the *Mimosa pigra* dilemma grows. While there have been no recent surveys of its present extent, its spread is encouraged by a number of new developments. In the past, it would infest only those river-banks which lacked tree cover (Plate 68). River-bank trees with vast interlocking root systems prevented erosion, and their wide canopies shaded the banks, denying the mimosa the favourable conditions it needed in order to thrive. With the imposition of the logging ban, grey market timber prices have risen substantially. Timber merchants along the Wang and other areas now offer irresistible prices to villagers for the shade trees growing along their banks. With the trees gone, the mimosa has multiplied, effectively fencing off the river, denying villagers cultivable area on the river-banks as well as access to fishing and hydroponic cultivation in the river itself.

68. Trees above the Wang River cut to permit the spread of the giant mimosa.

Dredging

Thais have always removed gravel from the rivers for local construction projects but in recent years sand dredging has become a large-scale commercial enterprise using powerful mechanical shovels (Colour Plate 27). At one operation on the Maenam Noi near Bang Sai, the river-bank has been cut back more than 1 kilometre from its original limits, creating an enormous lake. Whereas the Harbour Department scientifically dredges the lower river to facilitate navigation, the commercial companies remove sand without concern for its impact on portions of the river running from Chiang Dao to Bangkok.

Dredging licences are issued locally and often with minimal supervision of a company's operations. The ensuing intense siltation kills fish and changes river regimens, causing streams to shift and erode their banks. Farmers have protested but with minimal effect and there is little that individual landowners can do. If a landowner fails to agree to the company's price, the dredger trims around his land so he is left stranded on a small island that has a lowered property value and can only be reached by boat.

In some instances, the damage is more substantial. Dredging in the Chao Phya near Ang Thong caused the partial collapse of a river-bank downstream from the Chao Phya Dam.[11] On the basis of the report, the Harbour Department banned private dredging of the river, yet it continues. In a 1992 company newsletter, a major cement firm announced its intention to remove 200 000 tonnes of aggregate and sand from the Chao Phya near Ang Thong.

Overfishing

The damage to fish populations caused by pollution and siltation is complemented by overfishing which threatens to exhaust available supplies. Nearly all residents along waterways cast nets or lines for a variety of species for household consumption and commercial sale. Fishermen complain that today's fish are much smaller than in former times. It is likely that given the large number of people pursuing them the fish have little chance to grow to maturity. Their numbers are further depleted by the disappearance of migratory species blocked by dams—none of which have fish ladders—and by pollutants in the water. Fishermen often report the deaths of thousands of fish after heavy rain washes farm chemicals into the waterways.

Many fishermen employ techniques which kill large numbers of fish without discriminating between the fry and the fully grown. Dynamite, hand grenades, and pesticides poured directly into the water, seine-nets stretched the width of the river, and electric shocks are but a few of the illegal methods employed to catch fish which are then sold to unwary consumers in town markets. Many formerly abundant fish species such as *pla nam nguan, pla lin ma san*, and *pla lin mah yao* have disappeared altogether from the rivers.

Environmental Problems in the Lower Basin

Pollution is not new to the lower river. The simplest yet most pernicious form of pollution found along the river's length are the dead animals thrown into it. The bodies are snagged by shore vegetation and fallen trees where they putrefy, a situation vile enough to have once engaged the attention of King Mongkut. Seeking to remedy the situation, he penned a memorable edict in 1858 entitled 'His Majesty's Advice on the Inelegant Practice of Throwing Dead Animals into the Waterway, the Construction of Fireplaces, and the Manipulation of Window Wedges'. Mongkut understood that disposing of the 'carcasses of dead animals' posed a health threat as 'they float up and down in great abomination, and having thus contaminated the water, the City dwellers themselves do make an inelegant habit of constantly using the same water for purposes of drinking and ablution'. Having explained in clear detail why his subjects should cease the practice, he ordered them to bury the animals deep in the ground far from the city. In the future, anyone caught throwing a dead animal into the waterways would be paraded around the city in disgrace as an example to others for his/her 'inhumane and irresponsible act of water pollution'.[12] His order notwithstanding, the practice continues today, especially along the upper and middle rivers.

In 1889, writer Ernest Young noted the extreme degree of waterway pollution when he wryly commented, 'The water in the [curry vendor's] pot is drawn from the nearest canal or stagnant pool and is almost a meal in itself.'[13] Today, especially in the canals, the level of pollution from household waste has turned the water into an oily, semi-gelatinous, highly noxious soup.

The proposed Litchfield Town Plan of 1959 which included provisions for the construction of a comprehensive sewerage system was rejected for budgetary reasons. Since then, there has been much discussion of the need for sewage collection and treatment but, until recently, little of any substance had been achieved. The Public Health Ministry claims that only 2 per cent of Bangkok's buildings or houses are connected to a sewerage system. Most human waste is flushed into septic tanks, many of them faulty, which allow the germs to seep into the earth, contaminating groundwater and infiltrating residents' wells. In houses along the river and canal banks a majority of the toilets flush directly into the river.

Because the river is tide-affected, changing direction twice each day, much of the waste remains in the river for a considerable time rather than being swept out to sea. There, it rots and pollutes towns below Bangkok, with skin irritations and illness a common complaint, especially among children who swim in it. Below the city at Khlong Toey harbour, accidental (or deliberate) oil discharges from moored ships covers the river's surface in petroleum killing gastropods and fish eggs, and coating the mangrove leaves in a shiny black lacquer.

Extent of the Garbage Problem

In 1990, the Public Health Ministry estimated that Bangkok residents dispose of 180 000 kilograms of solid and liquid waste in the river and canals each day (Plate 69). The polluters are divided among residences (40 per cent), services (restaurants, fresh markets, hotels, guesthouses: 32 per cent), and industry (25 per cent). The Thailand Development Research Institute (TDRI), a semi-official development data and analysis organization, has calculated economic loss due to water pollution at B1,100 million ($44 million) annually.

In recent years, the levels of river contaminants have been quantified (Plate 70). In February 1993, the Public Health Ministry determined that the level of coliform bacteria, a key cause of cholera and diarrhoea, at the entrance to Khlong Phra Khanong was 1,362,500 MPN (Most Probable Number) per 100 millilitres, 68 times the Metropolitan Waterworks Authority (MWA) standard permissible level of 20,000 MPN and 27 times the standard of 50,000 MPN set elsewhere. Around Khlong Toey port, the Dissolved Oxygen level had dropped to zero, while in the river from Bangkok to Nakhon Sawan, it varied between 2 and 6 milligrams per litre; clean water normally contains 7 milligrams per litre. The UNCED Report on Thailand for the Earth Summit in June 1992 noted that 'discharges of domestic sewage and industrial waste to Bangkok's sewers, drains and waterways is estimated at 183,634 kg. of BOD/day'.[14]

69. A garbage dump on the banks of the Nan River above Phichit.

70. Debris on the shoreline at Phra Padaeng near the mouth of the Chao Phya.

Drinking Water

In the nineteenth century, many Bangkok residents obtained their drinking water from shallow wells. Most, however, drank directly from the river, resulting in illness and high death tolls, especially during plague years. To counter salinity in the river during the dry season and to halt the proliferation of pathogens, King Chulalongkorn in 1909, instructed the RID to build the city's first water treatment plant and piping system to supply potable water. Completed on 4 November 1914, it initially served 1,200 customers. Subsequent systems have drawn water from the Chao Phya just north of Bangkok, conveyed it along Khlong Prapha, and treated it at large plants. Most provincial towns now have water treatment facilities.

Flawed distribution pipes, however, allow leakage of large amounts of water. Although the water leaving the treatment plants is pure, the deteriorated pipes ensure that it issues from the tap partially contaminated by liquefied rust and by groundwater seeping into the rusted pipes. As a consequence, most residents boil their water before drinking it; the affluent drink bottled water drawn from deep-bore wells.

Deep-bore Wells

Historically, the delta's high water table ensured a wealth of aquifers in the clay-sandwiched layers of sand. Located a dozen metres below ground, the layers brimmed with water year-round and were replenished by surface water during monsoons and floods. In the 1920s, residents began boring deep wells and lining them with pipe. As factories were established, they, too, tapped the aquifers, withdrawing copious amounts of water, reducing their dependence on surface water from rivers and canals. In 1954, the MWA began pumping groundwater to augment its surface water supplies. Initially, it withdrew about 8000 cubic metres per day but by 1982 the daily total had risen to 447 000 cubic metres. In addition, hotels, factories, and housing estates were drawing about 944 000 cubic metres per day.[15]

To supplement water supply in the city and suburban areas outside the piping system, the Department of Mineral Resources (DMR) created the National Potable Water Scheme and by September 1985 had drilled 22,361 wells nation-wide, bringing the total for the country to 50,000–60,000 wells. Pumps extracted approximately 880 million cubic metres of groundwater each year, with about 475 million cubic metres consumed by domestic and industrial users.[16] As the city has grown, however, the sheer numbers of people have created dual but related problems. First, the aquifers have been seriously depleted, some to the point of extinction. Secondly, as a consequence of the first, the drained aquifers can no longer support the weight of the soils above them. As a result, between 1930 and 1990, Bangkok sank 1.7 metres;

80–85 centimetres of that fall occurred between 1978 and 1990.

The first signs of land subsidence appeared in the 1960s but by the 1970s it was reaching alarming proportions. Sidewalks were separating from buildings and bridges from their approaches. An Asian Institute of Technology study in 1978 revealed that in most areas of Bangkok, land was subsiding at a rate of more than 5 centimetres per year and in some places was sinking 10 centimetres per year.[17] By the 1980s, the suburban rate of decline equalled that of the inner city. Moreover, the subsidence in the metropolitan area and the eastern suburbs created a bowl. Trapped water could not drain over the bowl's lip and into the sea without extensive pumping.

To halt the subsidence, the government in 1977 passed the Groundwater Act banning the drilling of more wells. It also attempted to reduce the volume of water drawn from existing wells, but it took several years before the subsidence rate slowed. In addition, the MWA announced it would halt its water pumping by 1987 and support enforcement of the Act. Despite its resolution, it not only failed to cease pumping but in times of surface water shortage has increased the volume it extracts. For example, in November 1993, it announced it would pump 100 000 cubic metres of water per day to make up for shortfalls in tap water production, and the DMR would be asked to provide another 100 000 cubic metres in the same manner. In January 1994, the DMR announced it would bore 300 new wells.

Today, the situation has grown to serious proportions. According to the DMR in 1990, 1.2–1.4 million cubic metres of water per day were being extracted from 9,000 wells in the Bangkok Metropolitan Region. More than 80 per cent of this total was consumed by the private sector, mainly factories and housing estates. These figures, however, were based on known wells. A TDRI 1990 study deduced from independent estimates of industrial wastewater discharges that the true volume of extractions was nearly 3 million cubic metres per day.[18] If the TDRI is correct, then the problem is twice as serious as originally envisaged. As the upper aquifers have been drained, it has been necessary to bore even deeper. Whereas in 1968 the water table lay 12 metres below central Bangkok and 4 metres below the eastern suburbs, by 1993, it was necessary to drill 100–600 metres in central Bangkok to reach water.[19]

Since 1989, the subsidence rate in central and northern Bangkok has slowed to 1.5 centimetres per year as the upper aquifers have collapsed. In the eastern suburbs, it had been reduced to 7.0 centimetres per year, although by that time the area around Ramkamhaeng University had sunk to 30–40 centimetres below sea-level. The city continues to sink, abetting flooding and salinization, contaminating the aquifers, and necessitating substantial expenditure by the public and private sectors to repair damage to infrastructure, including roads and underground piping systems.

THE CHAO PHYA

More unsettling is the scenario predicted for the future if Bangkok continues to sink (Map 14). Studies in 1990 found that the sea had been rising 1.0–1.2 millimetres per year for the past 200 years. Global warming could raise sea-levels by 1.0–1.5 metres over the next 100 years. By the end of the twenty-first century, an accelerating rise caused by global warming combined with continuing land subsidence could see the creation of an enormous bay which would submerge Bangkok in a salt sea extending from Thonburi to the Hua Mak area and from the present Gulf (of Thailand) to Bang Khen near Don Muang Airport.[20] Changing weather patterns would also subject Thailand to tropical storms of the magnitude of those which strike the Philippines and Hong Kong each year.

Two related catastrophic events are occurring simultaneously. Delta growth is slowing because sediments which would normally build it are trapped behind dams, while other sediments flow into the river-bottom cavities created by the dredgers and never reach

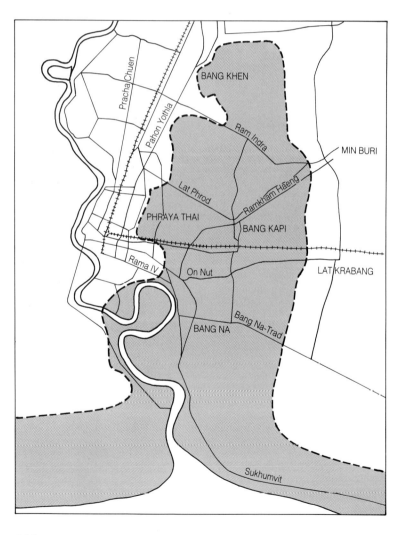

Map 14. The lower basin as it may look at the end of the twenty-first century.
(J. R. P. Somboon, 1990, p. 167)

the sea. Without the flow of aggregate into the Gulf, and with the destruction of mangroves which would buffer the shore from waves, oceanic forces erode the sea-shore faster than river-borne sediments can build it up. As has occurred in Taiwan, the vital margin between sea and land may simply disappear without expensive remedial measures such as placing thousands of concrete blocks on the shoreline to soften the wave action. The rise in sea-level will eventually eliminate the need for such measures.

It is clear that water extraction already exceeds natural replenishment. The first step in reducing depletions is to enforce regulations halting further drilling and to register illegal wells. In most cases, however, wellheads are small and can easily be hidden from inspectors. As these wells number in the thousands, identifying them could be a time-consuming process unless incentives are introduced to induce their owners to tie in to MWA piping systems. The MWA could also reduce pumping by raising rates to force users to cut consumption or improve efficiency by building treatment and recycling facilities. It would also require that the city find new sources of water to provide ample supplies to factories, specifically those whose technology demands enormous quantities of it.

Techniques to restore groundwater to its former levels include artificial recharge. During the monsoon rains, surface water is treated and then pumped down selected bore holes to replenish depleted aquifers. Economical and efficient, the approach has been successfully employed in Shanghai, Niigata, London, Paris, Germany, and the Netherlands. Studies in 1982 showed that water table levels could be raised by as much as 24 metres within two years.[21] Once new water treatment plants are operational, a recharging system would be a logical approach to ensuring future groundwater supplies.

Drainage

Despite the Greater Chao Phya Project, Bangkok continues to flood. Land subsidence has exacerbated the problem, rendering portions of the city lower than the river itself. At present, the city is partially protected by embankments raised by the Bangkok Metropolitan Authority on riverside streets. Water which overflows these barriers is removed by large mechanical pumps and returned to the main river. Recent proposals have ranged from constructing gigantic polders akin to those in the Netherlands, to a system of dikes and pumps, to large suburban reservoirs to store flood water for use during the dry season. Each scheme requires a substantial investment and the creation of a comprehensive water management master plan.

Water Weeds

Like the *mimosa pigra*, the water hyacinth is an imported weed. Struck by the beauty of its flowers, one of King Chulalongkorn's consorts returned from Indonesia with cuttings which were planted in ponds as decoratives. They eventually found their way into the canals and from thence into the rivers where they have become a hazard to navigation, entirely clogging canals or floating in enormous clusters down rivers during the monsoon season. In the 1930s, a Water Hyacinth Control Act was passed but it has had little effect in controlling its spread.

Programmes to turn the aquatic weed into an economic asset have been unsuccessful. Some companies have attempted to transform it into charcoal, others to turn it into farm compost, and still others have considered it as a source of wood pulp, but the initial investment costs are too high and the price of the end product is low. Self-help women's groups presently weave the dry stems into attractive cane for chair and sofa-backs but utilize a minuscule amount of it. The most plausible option is to harvest and bale it for animal fodder. Again, such an approach would likely be defeated by sheer volume and the speed at which the weed reproduces.

Aesthetics

On a question of pure aesthetics, the waterfronts which once enthralled foreign visitors have in many instances become riverside slums. While some provincial towns have turned their river-banks into popular promenades, Bangkok does not have a single waterfront park or even a landscaped quay on which to stroll. Plans for river-bank development range from the disastrous—a riverside expressway or tramway that would run 20 kilometres down the east bank and another that would run 15 kilometres down the western river-bank—to the enlightened: a landscaped park as an urban 'lung' connecting the green parklands of Ratburana and Pak Kret. None of the proposals has yet gathered the support required for implementation.

Despite laws prohibiting it, encroachment into the rivers is now common in large cities. Recent years have seen the construction of tall buildings along the banks of the Ping River, many in violation of city ordinances. Public protests have erupted over these structures, to no avail. A recent survey found that 79 per cent of a sample group of 500 Chiang Mai residents opposed construction of commercial or residential buildings along the Ping River; 47 per cent wanted recently built illegal buildings demolished. Still, construction continues.

Urban canals, which formerly served as Bangkok's streets, and the key components of its drainage systems, are dying. Most city canals have already been filled and concreted, providing roads but hampering drainage. Rather than being the pride of the city, the remaining canals are its cesspools, filled with oxygenless, oily sludge where no life exists. The Venice of the East has become the sewer of Asia. Instead of cleaning the waterways and policing the riverside residents who pour garbage and wastewater into them, the thrust of clean-up efforts has been on ridding them of boats. While cities like Paris and Amsterdam appreciate the aesthetic value of their boats and barges and have concentrated on ensuring that they do not pollute, municipal councils in riverside Thai towns have banned them altogether. The last houseboats/barges were removed from the Bangkok canals in the early 1980s, and Phitsanulok has announced its intention to remove the houseboats along its banks, citing their pollution and lack of toilet facilities as key reasons (Plate 71).

Loss of the Power of Taboos

In the past, taboos and spirits preserved the waterways. Today, they have lost their capacity to compel humans to respect the environment and desist from destroying it. 'In the domain of personal matters the spirits will maintain their powers.... It is in the domain of control of the natural elements that the spirits are no longer reliable.'[22] Farmers' reluctance to contribute to *muang fai* administration or pay fees or fines reflects, in part, a growing recognition that the old animistic strictures that once secured some measure of control within the system have lost their power. Education and exposure to the outside world—including labour experience in foreign countries—and a lack of knowledge of, or belief in, the system of spirits that govern *muang fai* have weakened the village taboos that rendered *muang fai* contracts sacrosanct and kept the weirs and channels safe from harm. Traditionally, annual offerings of pig's heads and foods were made to the weir spirit to ask its protection during the coming year. In 1981, in Phrae province, the few villagers attending one ceremony not only failed to make offerings but arrived with empty bowls to collect the foods that had been offered! Those few *muang fai* propitiation ceremonies which are still celebrated are presided over by young people and are often performed only to appease the elders who still believe in them.[23]

Similarly, watersheds are no longer protected except by a few hilltribes like the Karens. As on the river, traditional sanctions have failed to provide a deterrent to tree cutting. Phya Anuman Rajadhon noted that trees and wildlife were believed to be protected by *chao phi*, a superstitious belief which helped preserve them from destruction. In particular, the *yang* tree, a giant forest dipterocarp was considered sacred:

> Even today [1962] people in outlying districts will not dare to cut down a big tree for fear of the tree spirit residing in it. Even in felling a tree of smaller size, the people will first make an offering to the spirit to atone

71. At Phitsanulok, the last remaining houseboat community on any major Thai river is scheduled for removal in the near future.

for the offense made.... This was a wise practice to preserve big trees of the forest from wanton felling by the simple folk.[24]

In 1990, such constraints no longer dissuade villagers from cutting trees (Colour Plate 28).

With the loss of spiritual power as a controlling force, it has devolved upon humans to chart a new course to conserve their rivers. Numerous new questions are being asked and policies formulated, with 'sustainable development' the key term employed. Unfortunately, the concept is understood, or at least listened to, only by a handful of far-sighted technocrats, economists, and environmentalists who appreciate the folly of continuing along the present development path. To date, however, they have not prevailed over short-sighted developers and others who continue their depredations against the environment.

1. Phya Anuman Rajadhon, *Essays on Thai Folklore*, Bangkok: Social Science Association Press of Thailand, 1968, p. 195.

2. 'Conserve Water or Face the Consequences', *Bangkok Post*, 1 November 1992.

3. Ruangdej Srivardhana, *No Easy Management: Irrigation Development in the Chao Phya Basin, Thailand*, Reprint No. 63, Honolulu: East–West Environment and Policy Institute, 1984, p. 143a.

4. Nicolas Gervaise, *The Natural and Political History of the Kingdom of Siam*, Paris: Claude Barbin, 1688; English edn., trans. John Villiers, Bangkok: White Lotus, 1989, p. 14.

5. 'Out on a Limb', *Manager*, Bangkok, January 1992, p. 41, citing the Centre for the Suppression of Log Poaching and Forest Destruction of the Internal Security Operations Command of the Third Army Region.

6. Uraivan Tak-kim-yong, *Natural Resource Utilization and Management in the Mae Khan Basin: Intermediate Zone Crisis*, Chiang Mai: Faculty of Social Sciences, Chiang Mai University, 1988, p. 68.

7. Banpot Napompeth, 'Background, Threat, and Distribution of Mimosa Pigra L. in Thailand', in G. Lamar Robert and Dale H. Habeck (eds.), *Mimosa Pigra Management, Proceedings of an International Symposium, February 22–26, 1982, Chiang Mai*, Corvallis, Oregon: International Plant Protection Center, 1983, p. 19.

8. G. Lamar Robert, 'Economic Assessment of Mimosa Pigra in Thailand', in G. Lamar Robert and Dale H. Habeck (eds.), *Mimosa Pigra Management, Proceedings of an International Symposium, February 22–26, 1982, Chiang Mai*, Corvallis, Oregon: International Plant Protection Center, 1983, pp. 19, 28–30.

9. *Workshop on Uses and Losses due to Mimosa Pigra*, Chiang Mai: Chiang Mai University, 1986, pp. 87–8.

10. Interview with Banpot Napompeth, National Biological Control Research Center, Kasetsart University, Bangkok, 18 January 1993.

11. *Morphological Study of the Effects of Sand Dredging in the Chao Phya River, Thailand*, Research Report No. 96, Bangkok: Asian Institute of Technology, 1980.

12. MR Seni Pramoj and MR Kukrit Pramoj, *A King of Siam Speaks*, Bangkok: Siam Society, 1987, p. 53.

13. Ernest Young, *The Kingdom of the Yellow Robe: Being Sketches of the Domestic and Religious Rites and Ceremonies of the Siamese*, London: Archibald Constable & Co., 1898; reprinted Kuala Lumpur: Oxford University Press, 1982, p. 102.

14. *Thailand Country Report to the United Nations Conference on Environment and Development (UNCED), June 1992*, Ministry of Science, Technology and Industry, Bangkok, 1992, p. 36.

15. Anat Arbhabhirama, Dhira Phantumvanit, John Elkington, and Phaitoon Ingkasuwan, *Thailand: Natural Resources Profile*, Singapore: Oxford University Press, 1988, pp. 137–8.

16. Ibid., p. 104.

17. W. Piyasena and P. Nutalaya, 'Rehabilitation of a Depleted Aquifer System of Bangkok through Artificial Recharge', in *Soil, Geology and Landforms: Impact on Land Use Planning in Developing Countries. First International Symposium on Soil Geology and Landforms, Bangkok, 1–3 April 1982*, 1983, p. D9.2.

18. Sacha Sethaputra, Theodore Panayotou, and Vute Wangwacharakul, *Water Shortages: Managing Demand to Expand Supply*, Research Report No. 3, 1990 TDRI Year-end Conference, Jomtien, Bangkok: TDRI, 1990, p. 51.

19. Statistics supplied by the DMR.

20. J. R. P. Somboon, 'Coastal Geomorphic Response to Future Sea-level Rise and Its Implication for the Low-lying Areas of Bangkok Metropolis', *Tonan Ajia Kenkyu* [South-East Asian Studies], Tokyo, Vol. 28, No. 2, September 1990, pp. 155–68.

21. Piyasena and Nutalaya, 'Rehabilitation of a Depleted Aquifer System of Bangkok through Artificial Recharge', p. D9.1.

22. Richard P. Lando, 'The Spirits Aren't So Powerful Any More: Spirit Belief and Irrigation Organization in North Thailand', *Journal of the Siam Society*, Vol. 71, Pt. 1, 1983, pp. 121–47.

23. Ibid., pp. 141, 145.

24. Phya Anuman Rajadhon, 'Some Siamese Superstitions about Trees and Plants', in *Essays on Thai Folklore*, Bangkok: Social Science Association Press of Thailand, 1968, p. 304.

8 Shaping a New Role for the River

Modern irrigation schemes characteristically fail, or perform below their potential, because of the distance—both social and physical—between the irrigation engineers and managers, and the farmers they are deemed to serve.[1]

THE present state of the Chao Phya River system is precarious, and studies indicate that unless remedial measures are taken, the situation will deteriorate with disastrous consequences as Thailand approaches the twenty-first century. How will the future impact on the river system? What is being done to ensure that the Chao Phya River lives beyond the 2015 expiration date predicted for it?

In 1990, annual national water requirements for agricultural, domestic, and industrial purposes stood at 40, 2, and 1 billion cubic metres respectively. The Thailand Development Research Institute (TDRI) predicted a fourfold increase in water demand within two decades. It claims that in the year 2000, sectoral requirements would rise to 76, 6, and 3 billion cubic metres respectively. In 2010, they would stand at 144, 15, and 8 billion cubic metres respectively.[2] Yet in 1994, water shortages were already so severe that farmers were denied irrigation water for second crops and urban residents could expect tap-water supplies to dry up. It is clear that if a river management programme is not adopted soon, the North and the Central Plains face catastrophe as the Chao Phya River and its tributaries dry up, producing economic stagnation, and social disintegration with a sharp increase in the water rights battles like those which already plague some regions.

The principal impediments to ensuring that the river provides copious amounts of water are more human than hydrological, geological, or meteorological. Too many people and activities competing for the water, contradictory national development policies, uncoordinated water governance agencies, urban-based decision makers who take insufficient heed of grass roots concerns or expertise, inefficient and irresponsible water use, and major environmental abuses occur at every level of the economy and society.

Thailand's dilemma is alien to its experience. Historically awash in water, Thais believed that its rivers would always brim with lifegiving liquid. Thus, they became profligate in their use of it and are having difficulty grasping that the former largess no longer exists (Plate 72). Thus, taking the necessary conservation steps is ana-

72. Trying to draw water from a nearly dry canal.

thema, making implementation of water management more difficult than in desert countries which long ago embraced water conservation for sheer survival.

Thailand's water woes are not insoluble but overcoming them requires abandoning piecemeal approaches and the tendency to crisis manage. What is needed is a viable long-term land and water master plan which embraces and enriches everyone. More fundamental is the need to set priorities for national development in terms of water use—how Thailand's future will be balanced between industrial and agricultural pursuits—and to determine a new water allocation structure based on those priorities. It requires improving efficiencies in industry and agriculture to optimize use of limited water supplies, and producing goods which Thailand's natural resources and infrastructure are capable of supporting in a sustainable manner. Only if all else fails should it involve locating new sources of water. Even then, such appropriation should not compromise the environments and interests of inhabitants of other river valleys. Ideally, these policies would reverse the damage already done, and thereby improve the quality of life rather than simply boost per capita income.

Rethinking Objectives

In the past decade, most of Thailand's explosive economic growth has taken place in Bangkok, with industry making the most impressive gains. Although manufacturing has been targeted as the engine for growth into the twenty-first century, infrastructural inadequacies, social dislocation, and widening income disparity between rich and poor pose potential problems. The wisdom of rapidly industrializing a predominantly agrarian nation is debatable, but the concept is, and will continue to be, at the core of national economic development plans. It is vital, therefore, that planners pay close attention to the suitability and ecological sustainability of development ventures to avoid undermining the entire structure.

Water shortages are a major obstacle to continued prosperity, and a plague to planners. The area of greatest concern is agriculture which consumes 90 per cent of Thailand's available water resources, irrigating 2.9 million hectares with 40.32 billion cubic metres of water each year. A substantial portion of that total nourishes high-yielding rice varieties introduced in the 1950s and 1960s which made Thailand the world's premier rice exporter. Since rice's profitability is contingent upon world prices and because there is increasing competition from other grain-producing countries—many of which were formerly rice importers—Thailand's dominance as the pre-eminent producer is in jeopardy.

Moreover, it is not altogether clear that the conversion from low- to high-yielding rice varieties has benefited the farmer. In the past, he stored a two-year supply of rice in his barns and lived independent of the outside world. Now, the cost inputs for high-yielding rice (fertilizers, pesticides, the seeds themselves) require that he borrow money at the beginning of the planting season. He often must sell the rice before it is fully ripe in order to pay his creditors, many of whom charge exorbitant interest rates. Thus, he is in worse straits than he was thirty years ago. Providing the farmer with debt relief, farm credit at lower interest rates, or short-term subsidies would enable him to convert to less water-demanding rice strains or to new types of crop which would reduce his water consumption.

Since the nineteenth century, many Thai farmers have each year grown a second, very thirsty dry-season rice crop. In 1904, A. Cecil Carter noted that two crops were planted each year in the Central Plains:

... the first called 'Kao Bao' or light crop, and the second, 'Kao Nak' or heavy crop. The 'Kao Bao' is planted on irrigated land before the appearance of the rains in the plains, often as early as February, and is reaped in May or June. The 'Kao Nak' is planted between July and September, and is reaped in December or January. The 'Kao Bao' crop in no case amounts to a very large quantity of rice.[3]

With irrigation, the land devoted to Kao Bao has grown exponentially.

The appeal of dry- over wet-season rice to farmers in the North and Central Plains is founded on its substantially higher yields. For example, average yields for the 150-day dry crop during 1991/2 were 4368 kilograms per hectare, nearly double the 2643 kilograms for the 150-day wet-season rice. But, according to the Royal

Irrigation Department (RID), dry-season rice requires 1240 millimetres of water (wet-season rice needs only 799 millimetres) and its water requirements must be borne entirely by an irrigation system. Since wet-season rice constitutes a total of 3.6 million hectares and a dry-season second crop, 632 000 hectares,[4] a second crop in a decade of water shortages has become a luxury that Thailand can no longer afford. While it is unrealistic to return to an era when only rainfed broadcast upland wet-season rice was cultivated, it is necessary to halt the trend towards planting high-yielding rice varieties, and to eliminate the planting of second crops, especially on land newly brought under irrigation.

Moreover, given the present low world rice prices, it is worthwhile to ask whether Thailand should continue to place such heavy emphasis on rice exports or whether it should focus on other crops and development activities, and produce rice only for domestic consumption. Such a view, once considered heretical, is for the first time being discussed openly. In 1993, an Agriculture Extension Department official was quoted as saying that 'Thailand does not really have a comparative advantage in rice because the cost of the large consumption of water is not included in the price.... Thailand should only supply its own market and should forget exporting.' Were they to scale down large rice exports, economic planners would not be reversing a centuries-old tradition, but correcting a problem only one generation old, one that was created by the planners themselves.

Some steps in that direction have already been taken. Responding to the decline in rice prices, many Central Plains farmers have decreased the acreage devoted to rice cultivation. While rice acreage increased between 1976 and 1983, its portion of the total cultivated land area dropped from 61 to 52 per cent. Between 1984/5 and 1988/9, the total land area devoted to main season rice cultivation in the Chao Phya basin remained fairly steady, averaging 1 965 518 hectares per year. In the three years since then, it has declined to 1 685 785, 1 635 511, and 1 571 514 hectares respectively. Of these totals, the irrigated portion fell from 1 216 247 hectares in 1984/5 to 1 081 920 hectares in 1987/8, and then rose to 1 311 657 hectares in 1989/90. With the shortage of water, it declined again to 1 135 832 hectares in 1991/2 and 1 147 055 hectares in 1992/3. It was the non-irrigated portion that saw the largest change, rising from 747 498 hectares in 1984/5, to 864 800 hectares in 1987/8, then declining steadily to 499 679 hectares in 1991/2 and 424 466 hectares in 1992/3 as farmers converted to other crops.[5]

Farmers have adopted new strategies to cope with water shortages. Many have phased out the planting of 180-day rice, settling for 150- or 120-day rice, reducing water requirements by one-sixth and one-third respectively. Furthermore, farmers generally plant one species in one plot, and another variety in another plot in order to better weather water supply fluctuations.

The past two decades have also seen a change in plantation

profiles. For example, the total agricultural land devoted to field crops in the North rose from 23.0 to 35.6 per cent between 1976 and 1992 while that in the Central Plains rose from 21.0 to 32.9 per cent. In Central Thailand, fruit orchard plantation rose from 12.2 to 15.2 per cent between 1987 and 1992.[6] Maize, sorghum, cotton, mungbeans, soybeans and groundnuts, as well as some cassava, sugar, cotton, and tobacco, are the major field crops planted by northern farmers.

In the Central Plains, the Agriculture Ministry plans to promote mixed agrarian farming until the end of the twentieth century, reducing acreage for a second rice crop by planting legumes, maize, peanuts, and vegetables (Plates 73–75). While the objectives are laudable, the Greater Chao Phya Project, which was designed to serve rice cultivation, does not allow administrators to regulate water supply to the degree necessary to support the successful cultivation of field crops. In many instances, farmers lack the confidence or the financial resources to convert to new crops. However dismal the yields, they prefer the certainty of a rice harvest, three-to-four months after planting, to the prospect of waiting four-or-more years for a tree to bear fruit. Uncertain markets for many crops also fuel farmers' reluctance to switch.

In planning future water allocation, the increasing demand for electricity must also be considered. In the past thirty years, Thailand's energy requirements have risen from 1.6 million in 1961 to 18.0 million in 1983, to 35.1 tonnes of oil equivalent in 1992. Of the 1992 total, 70.68 per cent was produced in petroleum-fuelled thermal plants, 22.24 per cent in lignite-fuelled plants, 6.16 per cent by hydroelectric generation, and 0.92 per cent by other means. In 1993, annual electricity demand stood at 9730 megawatts and was expected to rise 10.30 per cent by 1996 and 7.76 per cent by the

73. Watering kale, which demands less water than rice.

74. Harvesting cabbages in a Thonburi truck garden.

year 2000. Thereafter, it would rise at 6.07 per cent per year until 2006 when it would stand at 25 515 megawatts. Of this 2006 total, 36.0 per cent will come from petroleum-fuelled generation plants, 3.9 per cent from hydroelectric generation, 21.5 per cent from lignite-powered plants, and 38.6 per cent from other sources including solar cells, wind, geothermal power, and purchases of electricity from Laos, Malaysia, China, and possibly other countries. In megawatt terms, the hydroelectric generation portion will total 4431.2 megawatts.[7] The Electricity Generating Authority of Thailand (EGAT) and the RID interpret this to mean that additional dams must be built to serve irrigation and electricity purposes, an endeavour rife with potential problems.

Reassessing Priorities

In order for the rivers to serve a broad range of needs, social groups, and income levels, the government must resolve several dichotomies.

Between Water Users Groups and Outsiders

The *muang fai* system as it now stands is inflexible, unable to expand to serve new rice areas or provide water for dry-season crops. The RID solution is to dam the streams at several points and divert the water into permanent trunk canals which feed it into existing *muang fai* channels for distribution to individual fields. While the system deprives the local *muang fai* administration of the authority to determine the amount of water it can remove from the river, it theoretically permits RID administrators to allocate water based on a national irrigation plan rather than serving one area to the detriment of another.

75. One of many types of water scoops used in truck gardens.

There is, however, a need to encourage farmer participation in planning and administering the *muang fai* and water users' associations so he has a vested interest in maintaining the canals. There is also a need for clearer guidelines to govern *muang fai* and users' association members as well as those farmers operating outside the associations' jurisdictions. Consultations, educational workshops, and dissemination of information on techniques and technologies can help the farmer feel he is involved in the process, not alien to it. Efforts must be made to incorporate within the structure those farmers who pump irrigation water directly from the rivers without consideration for other users. Such involvement can defuse conflicts before they escalate into confrontations.

Although farmers are essentially uncomfortable in opposing authority, they are self-directed, pursuing their own courses, often to the detriment of the system. As evidence, consider the government's inability for the past fifty years to collect water user fees. Farmers' reticence about discussing their objections, however, can lead to problems that government administrators learn about too late to resolve effectively or economically. Open lines of communication and a sense of commitment by all factions will contribute not only to rationalizing water allocation but also to ensuring harmonization of crop selection to avoid gluts and shortages.

Between Farmers of the North and the Central Plains

To date, development programmes have sought to boost crop yields by providing farmers with irrigation water. Project successes in the North (which holds 20 per cent of the nation's population) have led northern valley farmers to tap the rivers, removing such large quantities of water that little is left to flow downstream into reservoirs. This has deprived lower basin users of required supplies. In 1975, the irrigated area of the Ping basin was less than 150 000 hectares; in 1990, it stood at 300 000 hectares out of a total irrigatable area of 625 000 hectares. Reducing emphasis on rice grown for export, curtailing the number of second rice crops, and converting fields to alternative crops would lower northern water demands and boost reservoir inflows.

Between Rural and Urban Water Needs

As farm water requirements rise so, too, do those of a burgeoning urban population and the factories built on the outskirts of cities. The RID allows the Metropolitan Waterworks Authority (MWA) to withdraw 5.2 million cubic metres from the Chao Phya for Bangkok's daily water requirements. In the 1980s, the MWA estimated that, having phased out its groundwater extraction, it would need to draw 7.8 million cubic metres of water each day from the river by 2017. Yet, with only 6.0 million cubic metres flowing past the city each day, present levels of supply will suffice only until the year 2000.[8] If it is the government's intention to industrialize the economy then at some point it will have to deprive farms of river water in order to serve urban/industrial needs.

Between Irrigation and Electricity Needs

Although dams are touted as tri-purpose structures—irrigation water, electricity generation, and flood control—two of those objectives operate at cross-purposes: irrigation water and electricity generation. Of the total water stored in a reservoir, nearly two-thirds is reserved as 'dead storage', which cannot be utilized for agricultural purposes. 'Dead storage' serves two functions. It keeps the dam wet so it does not crack, and it raises the water to a height sufficient to develop a head whose pressure drives turbine blades to produce electricity. RID records in 1993 indicated that 'dead storage' water in eleven major dams totalled 23 260 million cubic metres including 3800 million cubic metres for the Bhumibol Dam and 2850 million cubic metres for the Sirikit Dam. That left only 12 648 million cubic metres of 'live storage' for irrigation and urban needs. In addition, although the times and amounts of water released are ostensibly set by the RID, copious amounts are discharged to serve EGAT's hydroelectric generation requirements during peak hours, usually at night when sleeping farmers are unable to utilize it for irrigation purposes. Considering that the water released at 6.00 p.m. can move at least 10 kilometres per hour—and that farmers generally do not begin work before 8.00 a.m.—the water travels 140 kilometres before it can be used.

One can justifiably ask whether urban and industry needs are being served to the detriment of the villages. To serve both sectors and to meet rising electricity demand it might be advisable to generate electricity by other means and reserve dams solely for farm and urban water requirements. If not, both interests might be better served by building smaller dams just downstream from main dams, to create 'compensation reservoirs', a technique employed in Syria and other countries to impound water released during electricity generation. Water flowing through the turbines at night would collect in the lower pond, and in the morning, would be released for irrigation purposes. A secondary dam is being built 1 kilometre below the Bhumibol Dam, not to hold water for later release but to store it so it may be pumped back upriver and into the reservoir for re-circulation through the electricity-generating turbines.

Between Competing Water Agencies

In 1993, water management in Thailand was overseen by three key organizations (the Office of the National Water Resources Committee, the National Rural Development Committee, and the National Environment Board) responsible for policy development, and governed by more than thirty agencies in seven ministries, and seventeen committees involved in various aspects of water provision or use. The duplication of effort, and overlapping areas of authority frustrates the creation and smooth implementation of a water management master plan. Streamlining and co-ordinating these agencies may not be realizable in an atmosphere where organizations and individuals have historically carved out bailiwicks and defended them fiercely, viewing external efforts to co-ordinate them as threats to their autonomy, authority, or even their continued existence. Implementing a comprehensive programme and encouraging co-operation and information sharing would require a great deal of patience and more than a measure of firm central government leadership.

Between Government and the Public

Two factors characterize public administration in Thailand: top-down decision-making and a highly centralized, Bangkok-based bureaucracy. In large part, the paternalistic management style is a historic relic dating from ancient times when the populace existed to serve the central government. Only recently has the concept of a democratic, responsive, and responsible government answerable to its constituents gained increasing support, at least among the emerging middle class. The traditional form of government, still well-entrenched, is characterized by a fortress mentality among planners who resist attempts by outsiders to obtain information, offer criticisms, or challenge in any way the dicta of the power élite. Decisions are often taken within closed circles without consultation with other agencies or with the grass roots affected by their decisions. As a result, conflicts and misunderstandings arise which compromise a programme's success. Many of these are couched in terms of farmer versus government or environmentalists versus technocrat/developers, and the rhetoric employed is increasingly strident.

Thus, while environmentalists were successful in halting construction of the Nam Choan Dam on the west branch (Khwae Yai) of the Mae Klong River which would have destroyed a national park, their triumph did not come about as a result of dialogue between planners and non-governmental organizations (NGOs), but because protesters were successful in rallying public opposition to the project. But with the proposed Pak Mun Dam in the Northeast, opponents were unable to persuade the government to accept the results of environmental impact studies. A similar battle looms over the proposed Kaeng Sua Ten project on the Yom River. It is revealing to note that the focus of the disputes has not been the potential damage to biodiversity or the environment but on government attempts to relocate villagers to new sites. This suggests that, unlike other countries, the environmental damage wrought by dam construction is comprehended by few people at any level in Thailand.

Such an adversarial confrontation is by no means confined to Thailand, as recent events in nearby countries demonstrate. For development programmes to benefit the widest spectrum of society, however, it is necessary that the government consider public participation and debate as a means of exploring a broad range of options rather than as a challenge to its omnipotent authority. If Thailand is to continue to regard itself as a democracy, then everyone must have access to basic data, and opponents to government programmes must be countered with persuasive arguments, not blunt force.

Within this context, it is also necessary to improve information networks. Most agencies are skilled at gathering information but often have difficulty locating it once it is in their files. There is no standardized system of recording data so that researchers find themselves comparing incompatible sets of statistics. A lot of the data are recorded on paper and are not analysed so that they are unretrievable or useless to all but the most dogged scholar. As Thailand advances towards a new information age, there is a pressing need to computerize data both to make them accessible and to enable analysts to interpret them.

Macro Responses

In its long history of providing water to the populace, the Thai government has progressed from the ancient expedient of siting farms and towns on river-banks, to channelling water from great distances (nineteenth-century canals), to creating irrigation areas (Rangsit and the Greater Chao Phya Project), to constructing large dams to capture water. Today, the social and economic costs of acquiring new sources of water as well as the alteration of, and the attendant environmental damage to, rivers and canals, necessitates the adoption of a demand side management approach to water allocation. This requires dividing available supplies among users in an equitable manner, and introducing water optimization programmes to conserve it.

In April 1993, EGAT adopted portions of a five-year demand side management master plan for electricity that anticipates saving 1070 gigawatt hours annually until 1996.[9] There is a need for a similar programme for water. The Thai government is now creating a Water Ministry to co-ordinate the activities of all water-concerned agencies. It is also preparing a master plan that would divide Thailand into twenty-five river basins. Basin administrators would compute the amount of water each user in the basin would receive and how much he would pay for it. The plan represents a welcome step towards decentralization, devolving authority for water allocation upon those authorities directly responsible to users.

In past decades, the Thai government has promulgated numerous water decrees but few reflect modern conditions, and even fewer are detailed enough to resolve disputes. Most lack clear guidelines and strictures, or penalties severe enough to deter violators. Recognizing the deficiencies of the present Irrigation Act, the government is currently drafting Thailand's first water bill, a comprehensive document that would supersede the Act and all ministerial decrees. The proposed law would authorize administrators to issue water permits and set water quotas. Such a system would facilitate the collection of water user fees and give downstream residents the option of suing upstream users for any problems they caused.

There is some doubt that the bill will reach legislators. A preliminary bill, an amalgam of drafts prepared by four different agencies, was to be submitted to Parliament in 1994 but progress towards its passage has been stymied by bureaucratic initiatives to create the Water Ministry. Critics contend that the government's aim is to forestall passage of the Water Law; instead of ruling by law, the ministry would simply dictate policy based on its own assessments of need.

It is hoped that if the water bill is enacted, the agencies concerned will be empowered (and willing) to enforce it for the common good and to be immune to internal or external coercion. As in other areas of Thai public administration, law enforcement can often be very selective, and is often characterized by a reluctance to punish wrongdoers. For example, in 1991, the government adopted a 'polluter pays' approach to environmental destruction. In 1993, a Khon Kaen sugar mill polluted four rivers in the North-east, killing thousands of fish, causing skin problems for riverside residents, and fouling tap water; officials described it as 'the country's worst ecological disaster'. Yet the perpetrators were excused on the grounds that they had not knowingly polluted but had extinguished a fire with settling pond water which then flowed into the river. No attempt was made by local authorities to discover why the mill did not have a properly functioning fire-fighting system. Similarly, the Industry Minister closed a river-polluting pulp and paper mill for a thirty-day investigation but refused to impose heavy penalties, saying, 'If we punish them, who will want to invest here?'[10] There is also a need to enforce logging laws to halt encroachment on watersheds and forests. By defining logging as a crime against national security, military units could patrol the forests to arrest poachers.

In a far-reaching decision, the government in 1992 recognized the role of NGOs in formulating sustainable development policies and in monitoring their implementation. NGO representatives are now invited as consultants to ministerial meetings. While some detractors still regard them as irresponsible gadflies, there is a growing consensus that they play important roles in decentralizing power and ensuring public participation in the policy formulation process. This is only a beginning, however. Active registered NGOs in all social and environmental areas number no more than two dozen (by comparison, the Philippines has more than 18,000) and many base their protests on emotional appeals rather than statistical evidence. Official encouragement could spur their numerical and substantive growth.

Improving Water Use Efficiency

Optimizing water usage will be the challenge of the coming decade and it will affect all sectors of the economy. In the countryside, it means improving irrigation efficiency from its present 15–30 per cent to the 50 per cent level achieved by many Asian countries. The difficulty lies in accomplishing it without building new irrigation infrastructure. A 1981 World Bank report noted that 80 per cent of Thailand's public investment in agriculture and 60 per cent of all public expenditure was for irrigation facilities. While water is a major factor in raising yields, investment in seeds, technology, and less-harmful fertilizers and pesticides could contribute significantly. Major improvements can be achieved simply by lining secondary and tertiary channels with concrete to prevent water loss during delivery to fields (Plate 76).[11] Another approach is to overturn centuries of water wastage by promoting water conser-

76. Irrigation efficiency can be improved dramatically by lining canals with concrete.

vation through a system of positive and negative incentives. One of the principal methods is to charge fees to rural and urban water consumers.

Farm water fees are perhaps the most controversial of all of Thailand's water policy issues. Van der Heide in 1903 recommended charging farmers for the water they used, as did the FAO in its 1948 feasibility study for the Greater Chao Phya Project. In each instance, planners sought to offset the costs of building and maintaining a national irrigation system. Then, as now, the imposition of user fees was rejected as a political issue. Although the 1942 Irrigation Act authorized the RID to collect up to 50 satang per *rai* (3.125 baht per hectare)—raised to B5 in 1975—political considerations forestalled its enforcement. Until the 1960s, the government regarded water supply as an official duty, a benevolence that stemmed from a monarchical tradition of providing for the needs of ordinary people. In the 1970s, water provision was seen as a tool to combat communism; were it to exact fees, the government would leave itself open to charges of exploiting the poor.[12] Since 1990, rising water production costs have revived the issue as the only practical means to ensure continuance of the irrigation system. The TDRI has stated the problem succinctly: 'The present shortage of water supply in many parts of Thailand demonstrates that the supply cannot catch up with the demand under current pricing and cost recovery policies.'[13]

There is an additional impetus to imposing water fees: to provide the farmer with an economic incentive to conserve his supplies. The problems inherent in a fee system are enormous, but they have been resolved in other countries. Farmers in China and other countries pay for the water they use and northern Thai farmers have been paying modest fees to maintain their *muang fai* systems. The RID supports the concept of user fees as do most development agencies. NGOs support them as long as the revenues are not used to build more dams. Concentration, they contend, should be on improving the efficiency of existing irrigation systems.

The constraints on the success of such a system are farmers' unwillingness to pay the fee, and legislators' reluctance to rouse their constituents' ire by supporting a fee structure. If an equitable system of water fees can be established and if the farmer can be guaranteed a reliable supply of water, he may be more willing to view it as simply another business expense. It has been suggested that farmer receptivity to water fees is contingent on the government's willingness to control the prices of agricultural commodities. The Commerce Ministry has also advocated crop price supports as a means to induce farmers to convert to less water-intensive field crops. Both proposals are unfeasible because much of the produce is exported and world prices cannot be fixed, as rubber, tin, and oil regulatory bodies have discovered. Moreover, a price subsidy programme—already demanded by many farmers—would be unacceptably expensive.

Devising a fee assessment and collection mechanism is a complex task. Fees can be based on the region of cultivation and type of crop, or can be based on simple volume measurements. The first is easier to administer but difficult to determine given the varying values of different crops. The second method involves installing water meters, a costly and not always reliable method.[14] It also requires personnel to monitor the meters, and to calculate and collect the fees. It might be more feasible to collect the fees at the time the crop is sold, deducting the amount from the sale price.

There are several alternatives. The TDRI suggests that farmers deriving more than 50 per cent of their income from agriculture be given a stake in irrigation systems, a system similar to the 'pollution credits' principle used in developing countries. At the end of the monsoon season, the RID would gauge the amount of water in storage and allocate it to each of the twenty-five river basin administrators. Regardless of whether he was in a rainfed or an irrigated area, a farmer would be allotted a portion of water which he could use, sell to other farmers, or sell to utility agencies. Thus, any water he conserved through improved irrigation efficiency or by converting to less water-intensive crops could provide him additional income.

Another approach is to reward users for utilizing equipment or techniques which improve water use efficiency. In many temperate countries, governments provide credits to individuals who insulate their homes or otherwise modify them to reduce energy consumption. Similar incentives could be extended to farmers for lining their irrigation channels or switching to rainfed crops. The credits could be in the form of grants to purchase the equipment, or water bill reductions in addition to the savings the farmer would enjoy by curtailing water use. The cost to the government of administering the programme would be offset by the savings derived by eliminating the need to develop new water sources. Such a programme could also give farmers an enhanced sense of involvement in their own user associations.

Parallel with farm water conservation incentives, higher water charges can encourage domestic, commercial, and industrial users in urban areas to be more water efficient. The MWA claims that present water rates are below production costs so that, in effect, it is subsidizing the water system and losing money. In 1994, Bangkok homeowners paid B4.00 ($0.16) per cubic metre for the first 30 cubic metres, rising incrementally to B7.65 per cubic metre after 201 cubic metres. Rates for offices were slightly higher. For industry, the rate started at B50 for the first 10 cubic metres, B7.20 from 51 to 60 cubic metres, and thereafter rose to a high of B8.60 for amounts from 201 to 2000 cubic metres. The rate then began to decline, dropping to B5 for amounts greater than 50 000 cubic metres. Undoubtedly, the lower rates are intended to encourage groundwater users to convert to surface water, but by making it cheaper to use more, rather than less, water, there is little incentive for industry to conserve it.

MWA studies suggest setting a base rate of B6.20 per cubic metre, the average rate paid overall for the volume of water normally used by a single household each month. Raising the rate to B9 per cubic metre would result in a decline in demand of 13 per cent (90 million cubic metres per month) and a revenue increase of B1.1 billion. A similar situation would be adopted for industrial users wherein a rise from a base price of B6.12 to B9.00 per cubic metre would result in a demand decrease of 23 per cent or 147 million cubic metres. Once again, however, the objections to raising the fees are political, with concern focused on the impact the increases would have on the urban poor.

It is also useful to ask how long such a fee incentive would remain effective. After the initial shock of a price rise subsides, users normally adjust to the increase and build it into their product prices. Thus, while a rate increase might be a deterrent for export products competing in an international market, it might be less so in a domestic economy that daily experiences cost of living rises for all categories of products and services. Furthermore, price rises might well lead to increased consumption of groundwater. One means of countering the latter would be to monitor wastewater discharges and measure them against claimed consumption figures, inspecting factories and fining those whose input and discharge figures did not tally. Another strategy would be to charge a manufacturer based on the amount of water normally used by a factory in that business category. Exemptions would be made if the factory owners could adequately document their source and consumption. As in rural areas, an urban credits system would reward industrialists and residents for installing water-saving devices such as low-water toilets, water-efficient taps, and other technology.

New Technologies

New technologies can conserve, optimize, and store water for future use. For example, with sub-surface drip techniques developed in Israel, underground piping systems deliver to a plant only the volume of water it requires, and shelter it against evaporation. The approach is especially effective in fruit orchards, or other areas where the soil does not require frequent tilling. The initial purchase and installation cost is high but the technique can be economical if water user fees are based on volume used. Most treated water contains pathogens that make it unsuitable for recycling in a city water system but it can be pumped to the countryside for use in drip irrigation. As the field pipes are buried 30–50 centimetres below ground, the possibility of soil contamination is low. The treated water also contains nutrients that enrich the soil, saving the farmer money on fertilizers.

Some approaches require a simple knowledge of science. Farmers water crops during the day, and although Thailand's humidity is high, a considerable quantity of water evaporates (Plate 77). Night

77. A mechanical sprayer operated during the day causes considerable water loss due to evaporation.

irrigation would reduce evaporation, allowing the water to soak into the ground rather than standing in the hot sun. While night farming is practised in many arid areas of the world, it is alien to the Thai farmer's experience and implementing it would require a major change of habits.

Water can be stored in reservoirs during the monsoon season for use during the dry season. More than 170 ponds have been constructed in the North-east since World War II. Depending on size, the shallow earthen ponds are designed to irrigate between 20 and 8000 hectares.[15] Unlike the ancient Khmer reservoirs which were sited over high water tables so that water not only did not leak into the subsoil, but was drawn into the pond by capillary action, Thai reservoirs suffer considerable water loss. Still, the Agriculture and Cooperatives Ministry plans to spend B600 million ($24 million) to construct 1,150 reservoirs in the Central Plains, each holding between 3000 and 5000 cubic metres. Presumably, they would be built over water tables or lined with concrete.

Ensuring proper water distribution may involve considerable expense to expand the watergate system currently in use on 25 per cent of Thailand's irrigation channels. One of two possible systems could be adopted. The 'structured' system requires that all the canals be full; the system lacks gates but farmers are informed when they are entitled to water and respond accordingly. The second system, the 'just on-time' approach, utilizes gates, thereby removing the need to keep the canals full at all times. Its drawback is that it requires constant monitoring to ensure that farmers do not manipulate the gates to gain more than their allotted share of water. The latter problem can be overcome by installing automatic feedback mechanisms monitored by computers. While this approach is more efficient, it is ultimately more costly to purchase and operate.[16]

Returning Rivers and Canals to Their Former Roles

In recent years, several attempts have been made to restore the rivers and canals to their former roles as transport channels. In 1979, the Harbour Department initiated a project financed by the World Bank to create wharves, loading, and storage facilities at Nakhon Sawan and at the Nan River town of Taphan Hin to reduce demands on road transport. Large barges would be berthed at the wharves to receive cargoes of gypsum, sand, bulk cargoes, bagged goods, and rice; a fleet of tugboats would push them to Bangkok. The success of the venture depended on the river being deep enough to float the barges. The Harbour Department secured an agreement from EGAT to release 80 cubic metres per second from Sirikit Dam, and 300 cubic metres per second from dams on the Ping and Pasak to ensure sufficient depth to move barges upriver beyond Chainat. Accordingly, engineers constructed an 80-metre-long wharf at Taphan Hin and a 120-metre one just below Nakhon Sawan. Godowns, maintenance workshops, and offices were erected and spacious receiving areas were surfaced in concrete.

By the completion of the project in 1988, however, upstream water used to irrigate second and third rice crops left insufficient supplies for the inland ports. After several studies, it was proposed that a dam be built downstream from Ayutthaya to raise the level of the Chao Phya. When studies revealed that such a dam would not affect levels on the Nan, the project was abandoned. Ironically, and perhaps an indication of the thrust of the future, the godowns were rented to a private firm as a transit point for rice transported by trucks from up-country farms to Bangkok.[17] The government has proposed building two more dams on the Chao Phya to raise water-levels for navigation and has asked the Harbour Department to conduct feasibility studies. Meanwhile, major bottling, glass, and cement companies now move products up and down the lower river by barge.

Since 1970, a river commuter service has sped passengers between Thanon Tok, just south of Bangkok, and the suburban town of Nonthaburi north of the city. In 1991, the profitability of the service inspired a second company to offer a route that runs from inside Khlong Bangkok Noi to Tha Chang Wang Luang at the Grand Palace, and upriver beyond Nonthaburi to Pakkret. On 1 October 1990, in response to growing road congestion, a canal ferry boat service was inaugurated on Khlong Saensap. Speedboats carry passengers between the eastern suburb of Bang Kapi and Phan Fa on Ratchadamnen Avenue, substantially reducing travel time. Despite the putrid odours emanating from the canals, the service proved so popular that in 1993, eight new routes were introduced, restoring the canals to a modicum of their former importance (Colour Plate 29).

New Attitudes Fostered

The 1990s have seen an important new development: concerted action to educate and motivate the public to preserve the environment by fostering awareness of a problem and the possible solutions to it. NGOs have been at the vanguard of education campaigns to raise consciousness regarding the Chao Phya River's pollution difficulties with 'Love the Chao Phya' poster contests, wall paintings, essays, and other programmes directed at schoolchildren (Colour Plate 30). Magic Eyes, Think Earth, and other non-profit organizations have focused urban children's attention on the river, initiating projects to use biodegradable materials in traditional products. A 1991 Magic Eyes campaign to use banana stalks instead of styrofoam as the bases for Loy Krathong floats resulted in the planting of tens of thousands of banana trees to provide fruit. As a result, only 3 per cent of the *krathong* bases that year were made of styrofoam. The same organizations have urged owners of riverside restaurants, markets, and businesses to find other means of disposing of garbage and wastewater but it is difficult to change entrenched attitudes and the river remains as filthy as ever.

Newspapers have invited advertising agencies to submit full-page advertisements on environmental themes, with prizes given to the best entries. Many other promotions by hotels, industries, and services stress their 'greenness' and commitment to the environment. Although there is a certain element of boarding a bandwagon, there is increasing awareness of the need to demonstrate social responsibility by addressing the topic.

Waste Disposal Techniques

Recognizing the pollution threat from garbage discarded on the river-banks, some small towns on the upper tributaries are experimenting with new approaches to waste disposal. The Nan River municipality of Chumsaeng has set aside a plot of land outside of town where garbage collected from town streets is left in the open for scrap dealers to retrieve metal and plastic bags. The garbage is covered every seven days, the normal incubation period for maggots. The method is rudimentary but it removes recyclable material, reduces the volume of garbage, and keeps most trash out of the rivers. In 1992, the Bangkok Metropolitan Authority (BMA) introduced a fleet of small boats to call thrice-weekly at canalside homes to collect garbage formerly thrown into the water.

Approximately 60 per cent of household water ends up as wastewater and with rapid urban growth, its collection and treatment has become a major challenge. Initial steps to create a sewer and treatment system that will eventually cover the entire city were taken in 1992 when the Office of the National Environment Board was upgraded to Cabinet-level status, and the B500 million ($20 million) Environmental Fund was created—later augmented by an infusion of B4,500 million ($180 million)—to finance construction of wastewater treatment plants in the Bangkok metropolitan area. BMA-contracted companies began work on five water treatment plants with a combined capacity of 255 000 cubic metres per day. This amount constitutes about one-quarter of the 1 million cubic metres of effluent discharged into the city's canals and river each day. Located at Rattanakosin Island, Yannawa, Talat Noi, Siphya, and Ratburana, the five plants were to be operational by late 1994. Other projects will follow. In Chiang Mai, construction has begun on a water treatment plant to serve 160,000 homes. The system will later be expanded to treat water from 300,000 homes in the city's suburbs.

It is estimated that 85 per cent of Thailand's industrial wastewater is highly polluted, with heavy metals only one of the many contaminants. Factories are now required by law to install wastewater treatment equipment. The National Environment Board's Water Quality Division estimates that while 80–90 per cent of the factories have water treatment facilities, only 60 per cent use them regularly. Two industrial wastewater treatment plants are being constructed with World Bank assistance at Rangsit and at Suksawat in Samut Prakarn province, one of Thailand's most polluted areas.

Since the 1970s, as part of his efforts to transform hill and valley agriculture, King Bhumibol has initiated hundreds of projects to impound small streams and village ponds for irrigation purposes. Photographs of him, accompanied by RID officials, striding along a rural waterway with a topographical map in his hand are in many respects the defining images of his reign. He has also designed water-wheel aerators to improve water quality in canals. Several have been placed in Bangkok's Makkasan Lake, and twenty-three are being installed in Chiang Mai's Mae Kha Canal.

Innovative Agriculture

As with the organic approaches to reducing mimosa, innovative techniques (Plate 78) are being tested to eliminate crop pests and reduce reliance on pesticides. *Phlia kradot si namtan* (Brown Leaf Hopper) annually infests rice crops in the Central Plains, notably along the lower Nan River where, in 1990 and 1991, it destroyed major portions of the rice crop. Insect larvae are killed by long immersion in water, as noted by la Loubère when he wrote: 'Besides the Inundations fatning the Land, it destroys the Insects.'[18] Unfortunately, the river cannot always be depended upon to rise at the appropriate time. Researchers at an Agriculture and Cooperatives Ministry research station at Chainat placed *chamung thanu* (archer fish, *Toxotes Chatareus*) in a test plot of rice. The fish can squirt a stream of water 1.5 metres, knocking the hoppers off rice plants and then devouring them. In experiments, a half-dozen fish cleared a large plot of rice hoppers in less than one week. Were a widescale project to be implemented, the fish would be placed in the

THE CHAO PHYA

78. Frog farming is becoming popular in villages along the river-banks.

79. These melons, under their polyethylene protection, receive only the amount of water they require.

rice fields at the beginning of the planting season and collected by farmers as a protein supplement at harvest time.

To curtail pesticide use, many farmers in the Chiang Mai Valley now cover field crops with mesh netting. Others stretch sheets of reflective plastic sheetings over beds of melon seedlings which protrude through small slits in the plastic (Plate 79). The silver sheeting reflects most of the sun's heat, inhibiting evaporation and denying weeds the sunlight they need for photosynthesis, as well as eliminating the need for herbicides.

Integrated farming could supply compost to replace harmful chemical fertilizers now in use. Rice fixes much of its own nitrogen so applications of manufactured fertilizers only augment nature's own supplies. The replacement of water-buffaloes by rototillers (Colour Plate 31) has reduced the amount of animal manure available for compost but farm families maintain cows and pigs whose manure could be used. There have been numerous successful experiments to make low-cost biogas converters which utilize animal droppings (unlike China, there is a repugnance towards using human waste), kitchen waste, and vegetable matter as fuel to produce methane gas for kitchen use, and compost for gardens. A large-scale programme would reduce the amount of chemical fertilizers now polluting the streams and rivers.

Land and Forest Management

The era when Thailand had ample land reserves for future expansion has ended. Fortunately, a number of encouraging developments have taken place in the past decade which will reduce the pressure on the land. The rampant population growth of the 1970s has been curbed through highly successful family planning programmes, and urban employment is absorbing young people who might otherwise be clearing watershed forests for cultivation. Both of these, however, are insufficient to halt further degradation. Laws against encroachment must be enforced and loopholes, such as the 'degraded forests' category which allows unscrupulous officials to declare an area ripe for commercial replanting, must be closed.

Farmers must be taught to make more efficient use of existing land, improving rice yields rather than expanding their holdings (Colour Plate 32). Thailand's harvests per hectare are among the lowest in Asia, less than one-third those achieved by many countries (Table 6). In part, they are low because Thai farmers have never been forced by land scarcity to practise intensive cultivation; with ample supplies of land, they had little need to exert themselves. As a result, although they accepted new high-yielding rice varieties, they did not adopt the dozen or so associated processes required to make the rice species highly productive. Many of these techniques involved regulating the water-levels at different intervals during the growing season. Coupled with their reluctance was an irrigation system which did not permit fine tuning. With current land pressures and low commodity prices, it will soon be necessary for the irrigation system to be altered to boost yields. Making the land more productive would lower the demand for water and reduce the need to cut forests.

Restoring watersheds has become the new imperative of forestry management. The Prime Minister's Office has embarked on a massive tree-planting scheme along the upper reaches of the Ping, Wang, Yom, and Nan Rivers. The programme, which has encountered considerable opposition, involves relocating nearly one million people, 90 per cent of them tribal, from 3,500 villages which have been illegally established in the watershed areas. More than 160 000 hectares of forest will be planted by 1997. Finding

TABLE 6
Comparative Rice Yields, 1990 (kilograms/hectare)

Country/Area	Yield
World average	3568
Asian average	3650
Japan	6325
China	5725
Indonesia	4318
Vietnam	3118
Philippines	2806
India	2693
Nepal	2306
Pakistan	2218
Thailand	1956
Cambodia	1331

Sources: Thailand: Office of Agricultural Economics; other countries: *FAO Production Yearbook*, 1988, 1989, 1990.

new homes for this enormous number of displaced people in an area already short of arable land poses a considerable problem, especially when recognizing that they migrated to that region because they could not find suitable farmland elsewhere.

Forestry officials and Chiang Mai University academics have been studying traditional Karen hilltribe conservation methods to learn techniques and philosophies which could be applied in other tribal and Thai upland villages. The programmes are successful only on the micro-level, but as competition between the tribes and upper valley farmers escalates, mutual management schemes will become imperative; the Karen experience could facilitate their implementation. Karen villages in Ban Mae Om-Long Klang of the Hot District of Chiang Mai province, for example, have strict rules against felling timber in watershed areas. They also practise communal forest management planting green belts around the villages to permit natural regeneration of cleared areas. Fire-breaks are dug between cultivated fields and forests to halt the spread of accidental fires.[19]

Dams and Diversions

Since the 1960s, His Majesty King Bhumibol has sponsored rain-making efforts, maintaining a small fleet of aeroplanes to seed clouds during the monsoon season. In 1993, the Royal Rain-making Research and Development Institute seeded clouds over dam reservoirs with 5–10 tonnes of simple chemical compounds—sodium chloride, urea, and ammonium nitrate—and reported that it had 'helped' to increase water flow into the Bhumibol Dam from 4 million to 34 million cubic metres a day. As the statement was made during September, traditionally the month of greatest rainfall, it is difficult to gauge its validity. Since water-levels have been so low for so long despite such efforts, it would appear that the rain-makers' contribution is negligible.

The failure of the skies to provide water has led planners to turn their eyes from the skies to the ground. Emphasis on better management of present supplies is regarded by the RID as having limited value. A RID spokesman was quoted as saying, 'We don't know how much water will be retrieved from water-saving campaigns and an adjustment of farming methods, but we are sure that it won't be a large amount.'[20] Convinced that conservation methods will be insufficient to secure adequate supplies of water for future needs, the Thai government is, once again, turning to hydraulic engineers, touting dams and diversion projects as the vehicles of deliverance from chronic water shortages.

The RID would like to build four more major dams in the North:
1. Sakae Krang River, Uthai Thani province, a tributary which enters the Chao Phya from the west between Phayuha Khiri and Manoram; it has a potential capacity of 1 billion cubic metres of water.
2. Yom River at the Kaeng Sua Ten rapids above Phrae; its potential capacity is 1.2 billion cubic metres.

3. Pasak River, at Kaeng Khoi in Saraburi province or Phattana Nikhom in Lopburi with a potential capacity of 1.5 billion cubic metres.
4. Nan River, in Phitsanulok with a potential capacity of 1.3 billion cubic metres.[21]

Dams are also being proposed for hydroelectric purposes. EGAT estimates Thailand's total hydropower potential, excluding international projects, to be 10 626.2 megawatts, four times the present generating capacity of 2416 megawatts. New generating units added to existing powerhouses can expand production by 300 megawatts, but after that, it contends, new hydroelectric dams must be built. EGAT proposes to construct six hydroelectric dams that would raise the generating capacity by 1702 megawatts by the year 2003.[22]

In support of its proposals, the RID claims that only 17 per cent of Thailand's rainfall is impounded in dam reservoirs; 83 per cent flows unused into the sea. Of the 24.0 million hectares of land presently under cultivation, only 3.3 million hectares or 14 per cent is irrigated. Dams and diversion projects would bring irrigation to the remainder.

Critics contend that it is useless to talk of water running into the sea unutilized. To begin with, a major portion of the runoff is in the lower basin where it is impossible to build dams which can impound water efficiently; the gradient is so low that a thin layer of water would be spread over a wide area, inundating huge tracts of fertile land already under intensive cultivation. In essence, the water is already being impounded by the bunds or low walls that hold it within individual rice fields. They contend, furthermore, that in the North, there are numerous dams to trap monsoon runoff. If there is excess flow, they ask, why are reservoir levels so low?

Building new dams and reservoirs in the North is impractical because most of the suitable sites have already been used. Once the few remaining sites have been utilized, the government will be forced to resort to the only alternative available: reconsidering priorities, husbanding supplies, and drafting new water use and economic development policies that redirect the country away from heavy dependence on water. Rather than waiting until the situation reaches crisis point, observers advise planners to shelve dam projects, concentrate on improving irrigation and water use efficiency, and restructure agencies and activities to achieve optimum co-operation and co-ordination.

TDRI analysts oppose dams on economic grounds. They note that fluctuating international rice prices, the land requirements for reservoirs given the scarcity of fertile land, and the investment costs for large-scale projects make dam construction uneconomical. Moreover, only 20 per cent of Thailand's arable land is irrigatable. Instead, it advocates 'making better use of the potential of existing projects by means of modernization and/or rehabilitation of the infrastructure'.[23] Observing the damage caused by dams, Wildlife Fund Thailand (WFT), an NGO, urges that greater effort be spent on replanting watersheds which will act as natural, more efficient means to trap and release water. Instead of flowing over denuded areas in flash floods as it now does, surface water would be absorbed, and then released slowly over a long period. According to the WFT, 'the eight years spent on constructing a dam and the two to three years needed to store the water is a period in which a forest can be recovered'.[24]

One potential dam site is opposed by environmental groups. The Kaeng Sua Ten Dam, proposed for the Yom River, would store 1.175 billion cubic metres of water in a reservoir covering approximately 66 square kilometres. The project would be built in two phases. In the first, a 72-metre-tall dam would impound water to irrigate 50 000 hectares in the Yom basin. Its height would later be raised to 96 metres to power a hydroelectric generating plant of 230-megawatt capacity. A portion of the water would be diverted to the Nan River to augment flows into the Sirikit Dam reservoir. A World Bank-financed Mahidol University environmental study of the dam site, however, noted that it is in the middle of the Mae Yom National Park and more than 4800 hectares of golden teak forest, considered to be 'the richest teak forest remaining in Thailand', would be inundated. Land and water species migrations would be interdicted by the reservoir, reducing population numbers and threatening biodiversity in one of the few areas of Thailand containing such a wealth of animal and plant species.[25] Despite the report, the RID continues to press for the project.

Existing reservoirs must also be better utilized. Of Thailand's twenty-five major dams, fifteen are storing water at less than 50 per cent capacity. One problem is sediment accumulation to a degree not anticipated when the dams were first constructed. The sedimentation rate has been exacerbated by the clearance of forests which has accelerated the erosion of topsoils, increasing the sediment content of the Bhumibol Dam reservoir by 9 million cubic metres per year. For the thirty years since the construction of the dam, this translates into a total deposit of 270 million cubic metres, an amount that will considerably shorten the dam's effective life. Removing the sediment would create a larger storage area. That can be accomplished at great expense by suctioning out the silt or could be achieved by placing sluice gates in the lower dam and letting the silt flow out with the escaping water, a not inconsiderable expense but more affordable than replacing the dam long before its intended expiration date.

New dams are but one component of a grand scheme to increase Thailand's water supply. The other is to divert portions of Thai rivers, as well as rivers in neighbouring countries, into the Chao Phya and its tributaries.

The first major diversion proposal, called Chao Phya II, was designed to drain water away from, rather than towards, the Chao Phya River. Intended as an anti-flood measure for Bangkok, it was initiated in the aftermath of the disastrous floods of 1983. It would have involved digging a canal 60 kilometres long, 220 metres wide, and 10 metres deep to channel the Chao Phya's water into the Mae Klong River west of Bangkok. An alternative was to divert the Pasak River into the Bangpakong River and thereby reduce the amount of water flowing into the Chao Phya at Ayutthaya. By 1988, faced with potential water shortages, the emphasis shifted from redirecting water away from the Chao Phya to diverting rivers into it.

Studies reveal that 50 cubic metres per second of excess water in the Mae Klong River can be channelled into the Chao Phya. According to the 1988 plan, the diverted water, approximately the amount that the Chao Phya must pour into the Tha Chin at Manoram to counter excessive salinity in that river, would provide dry-season irrigation for approximately 50 000 hectares. According to a revised plan approved in 1994, the water will be used by Bangkok residents for drinking purposes. A 106-kilometre canal will carry water from Vajiralongkorn Dam to a treatment plant to be built in Bangkok; the entire project will be completed in 1998. In 1993, the RID announced it would like to add the Bangpakong River, east of Bangkok near Chon Buri, to the scheme, channelling its waters into the Chao Phya. In addition, the Pai River in Mae Hong Son and the Moei River in Tak would be diverted into the Ping River. Already, however, the proposed scheme is encountering opposition. A Kasetsart University study has noted that by the year 2000, farmers in the Mae Klong Valley will need 9400 million cubic metres of water from the river. Diverting a portion of the river to Bangkok would deprive Mae Klong farmers of irrigation water.

An ancillary but vital component of the Kaeng Sua Ten Dam project is the Ing–Yom–Nan Water Diversion Project, which was first considered in a 1982 EGAT study. The gigantic eight-year scheme would pump portions of two Mekong tributaries, the Ing and Kok Rivers, south into the Kaeng Sua Ten reservoir. At the same time, some of the water from the Kaeng Sua Ten reservoir would be pumped into the reservoir of the Sirikit Dam. More ambitious is a plan to divert a portion of the Mekong River, which flows through six nations, into the upper Nan Valley. This would be feasible only if the taller of the two designs for the Pamong Dam were built on the Mekong, a proposal which has already been rejected because of the more than one million villagers who would have to be relocated. Also under consideration is a plan to buy Salween River water from Burma and channel it into the Ping River, an enormous undertaking which would require lifting it over the mountain range that forms the country's western border.

The diversion schemes raise an important question: To whom does a river and its water belong? Do the people living along its banks have a claim on it or can a national authority determine its end user? According to the drafts for the Water Law, all rivers belong to the central government. As the smaller valleys develop,

THE CHAO PHYA

however, they will have their own water needs and a choice will have to be made as to whether they or the urban areas have first call on the rivers' waters. Worse yet, if the cities have become dependent on the diverted supplies, they will have to curtail consumption when the original owners reclaim their rivers. Once again, the diversion approach is short-sighted, solving an immediate problem without reference to future requirements. It also suggests sacrificing the interests of the smaller valleys to serve those of the Central Plains.

On an international scale, the TDRI suggests that it is economical to buy water from adjacent countries because of the current 'co-operative atmosphere' between Thailand and its neighbours. Laos has offered to sell water from its rivers, piping it under the Mekong River to avoid international repercussions with other riparian nations. The water would be pumped into Thailand's northern and north-eastern river systems.[26] While it is true that Thailand annually buys 146.7 megawatts of electricity from Laos and 30.6 megawatts from Malaysia—the 177.3 megawatt total represents 1.82 per cent of Thailand's supply—the electricity is sold as an export product. With water, a physical entity is transferred. In the case of Burma's Salween River, Thailand's dependence on a foreign source means radically revising its own system if the neighbour later decrees its own need for the river and rescinds the contracts, or if a dispute arises between the two nations. Contrary to the assertion that there is a 'co-operative atmosphere', Thailand is being criticized for exploiting its neighbours' natural resources. It seems more important that Thailand solve its water dilemma now rather than in the future when a sudden cessation of supplies could create larger problems.

Among the other alternatives being considered in order to augment supplies is to extract groundwater from up-country sources and pump it into the Chao Phya River (Plate 80). Two experimental groundwater development projects are operating in Sukhothai and Phichit to supply water for dry-season agricultural purposes but have not yet proven to be feasible because of the high cost of exploration and piping systems, and the small size of the aquifers. It is evident that much work remains to be done, but it must be considered as a single component of a larger, all-embracing programme. At the same time, the Science, Technology, and Environment Ministry has allocated B500 million to dig 50,000 shallow wells throughout the country. The Civil Disaster Prevention Committee is also financing the boring of 3,380 artesian wells in 6,242 villages in drought areas to provide water during the dry season. While some Central Plains villagers have used existing wells to irrigate up to 0.5 hectares of rice fields during droughts, it is insufficient to serve large-scale irrigation needs.

80. Searching for new sources of water in an increasingly arid land.

Contemplating the Future

When contemplating the Chao Phya's placid surface as it rolls through the heart of Thailand, one is aware only of its power and its magnitude. The farms and towns it sweeps past owe their existence to its life-giving waters, just as the ancient kingdoms brooding on its banks owed it their glory. Explorations of its past reveal its central role in shaping history, economies, values, and cultures; a rich contribution to human life that poet Ted Hughes calls the 'vintage [of] unending river [that] swells from the press of earth'.[27]

In its pettish moods, its ravages can be as extreme as its blessings, but they are minor aberrations in Mae Khongkha's otherwise gentle mien. Once they have passed, she returns to nurturing those whom she serves. But the goddess's patience is being tried. From even a cursory examination of its role and its state today, it is clear that the once-mighty river system faces an uncertain future. From its omniscience as the nation's lifeline, it has been reduced to a slave of development, exploited and abused in myriad ways. It is clear that it cannot long endure such treatment. With clearly defined and unified priorities and objectives, the river can be returned to the vital role it once served. At the moment, however, the future of the Chao Phya River holds more problems than promise. Resolving them is the task of the decade, and the decade is half gone.

---

1. Jonathan Rigg, 'The Gift of Water', in Jonathan Rigg (ed.), *The Gift of Water: Water Management, Cosmology and the State in South East Asia*, London: School of Oriental and African Studies, University of London, 1992, p. 5.

2. Sacha Sethaputra, Theodore Panayotou, and Vute Wangwacharakul, *Water Shortages: Managing Demand to Expand Supply*, Research Report No. 3, 1990 TDRI Year-end Conference, Jomtien, Bangkok: TDRI, 1990, p. 6.

3. A. Cecil Carter (ed.), *The Kingdom of Siam 1904*, New York: G. P. Putnam's Sons, 1904; reprinted Bangkok: Siam Society, 1988, p. 160.

4. Center for Agricultural Statistics, *Agricultural Statistics of Thailand, Crop Year 1991/92*, Office of Agricultural Economics, Ministry of Agriculture and Cooperatives, Bangkok, 1992, pp. 12–13, 18–19.

5. Ibid., pp. 212–13.

6. Ibid.

7. Electricity Generating Authority of Thailand, November 1993.

8. Sacha, Theodore, and Vute, *Water Shortages*, p. 63.

9. *Demand Side Management for Thailand's Electric Power System; Five-Year Master Plan*, Prepared by the International Institute for Energy Conservation, Bangkok, 1991.

10. 'Sanan: No Need to Get Tough with Factories', *Bangkok Post*, 29 May 1993.

11. Ruangdej Srivardhana, *No Easy Management: Irrigation Development in the Chao Phya Basin, Thailand*, Reprint No. 63, Honolulu: East–West Environment and Policy Institute, 1984, p. 143.

12. C. L. J. Van der Meer, *Rural Development in Northern Thailand: An Interpretation and Analysis*, Groningen: Krips Repro Meppel, University of Groningen, 1981, p. 190.

13. Sacha, Theodore, and Vute, *Water Shortages*, p. 59.

14. Anat Arbhabhirama, Dhira Phantumvanit, John Elkington, and Phaitoon Ingkasuwan, *Thailand: Natural Resources Profile*, Singapore: Oxford University Press, 1988, p. 128.

15. Yoshihiro Kaida, 'Irrigation and Drainage, Present and Future', in Yoneo Ishii (ed.), *Thailand: A Rice- growing Society*, trans. Peter Hawkes and Stephanie Hawkes, Honolulu: University Press of Hawaii, 1978, p. 223.

16. Scott R. Christensen and Areeya Boon-long, *Institutional Problems in Thai Water Management*, Bangkok: Thailand Development Research Institute, 1993, pp. 18–19.

17. *Project Completion Report of the First Inland Waterway Project*, Bangkok: United Nations, 1990.

18. Simon de la Loubère, *A New Historical Relation of the Kingdom of Siam*, London, 1693; reprinted Kuala Lumpur and Singapore: Oxford University Press, 1969 and 1986, p. 15.

19. Uraivan Tan-kim-yong, *The Karen Culture: A Co-existence of Two Forest Conservation Systems*, Chiang Mai: Faculty of Social Sciences, Chiang Mai University, 1990.

20. 'The Official Answer: Dam Projects', *Bangkok Post*, 3 November 1992.

21. Royal Irrigation Department Statistics, 1993.

22. Electricity Generating Authority of Thailand, November 1993.

23. Anat, Dhira, Elkington, and Phaitoon, *Thailand: Natural Resources Profile*, p. 127.

24. 'The Official Answer: Dam Projects', *Bangkok Post*, 3 November 1992.

25. Center for Conservation Biology, *Rapid Assessment of Forest/Wildlife/River Ecology in Area Affected by Kaeng Sua Ten Dam*, Bangkok: Mahidol University, 1992.

26. Sacha, Theodore, and Vute, *Water Shortages*, pp. 78–9.

27. Ted Hughes, 'The Vintage of River Is Unending', *River*, London: Faber and Faber, 1983, p. 66.

Glossary

Baht, a unit of Thai currency, in 1994 valued at B25 to US$1.
Bot, the hall in a *wat* reserved for the ordination of monks.
Chedi, a tall spire originally raised over relics of the Buddha but now used to hold ashes of a deceased Buddhist.
Doi, a northern word for 'mountain'.
Hua na fai, the village irrigation official responsible for administration and maintenance of the *muang fai* system; earlier named *khun nai fai*.
Huai, a northern Thai word for 'stream'.
Kamnan, a subdistrict headman, normally elected by a village or several villages.
Khlong, a canal.
Khlong chuam maenam, a canal designed to link one river with another.
Khlong lat, a canal excavated across the oxbow of a meandering river to shorten travel distances.
Khlong rop muang, a canal around a city, generally excavated as a protective moat.
Khwae, a branch of a river.
Krathong, small boats, generally made of organic materials, laden with candles and incense sticks and launched into waterways on Loy Krathong evening to seek blessings and ask forgiveness from the river for sins committed against it.
Lam, a northern Thai word for 'stream'.
Maenam, translated strictly as 'Mother of Waters' but generally as 'river' and applied to major streams.
Mon or *Mon–Khmer*, a people originating in southern Burma who began moving east sometime before the sixth century, settling in the lower Chao Phya Valley or moving further east to become the dominant culture that ultimately created the Angkorian civilization of Cambodia.
Muang fai, a locally designed and built irrigation system comprising small dams and channels to water crops; found on northern streams.
Nak, the Thai translation of the Sanskrit term *naga* and describing a mythical serpent associated with water.
Nam, strictly translated, it means 'water' but also refers to a small stream.
Nam mae, indicates a stream slightly larger than a *nam*.
Phu yai ban, the headman of a village.
Rai, a unit of area equal to 0.4 acres.
Rua, a boat.
Rua khut, a dugout carved from a single log.
Rua mat serm krap, a dugout with gunwales.
Rua taw, boats assembled from planks.
Tai, progenitors of the Thais and the Thai Yai and other cultures of a related language group; they appear to have originated in southern China and spread west into Burma and east into northern Vietnam.

GLOSSARY

Thai Yai, a branch of the Tai found primarily in north-eastern Burma where they are known as Shan.

Wa, a unit of length equal to 2 metres.

Wat, a Buddhist monastery; the term is often translated as 'temple' but describes a complex which can encompass several *wihan*, a *bot*, monks' quarters, meeting halls, and other buildings reserved for religious purposes.

Wihan, the worship hall of a *wat*.

Bibliography

Allersma, E., Hoekstra, A. J., and Bijker, E. W., *Transport Patterns in the Chao Phya Estuary*, Publication No. 47, Delft: Delft Hydraulics Laboratory, 1967.

Anat Arbhabhirama, Dhira Phantumvanit, John Elkington, and Phaitoon Ingkasuwan, *Thailand: Natural Resources Profile*, Singapore: Oxford University Press, 1988.

Ancient Customary Laws and History Laws of the North, Microfilms of palm leaf manuscripts dating from King Mengrai's reign, Chiang Mai University library.

Anuman Rajadhon, Phya, *Chiwit Chao Thai Samai Kon le Kan Suksa Ruang Prapheni Thai* [Thai Life in Olden Times and Studies of Thai Traditions], Bangkok: Samnak Phim Khlang Wittaya, 1972, pp. 301–10.

_____, *Essays on Thai Folklore*, Bangkok: Social Science Association Press of Thailand, 1968.

Ayutthaya Chronicles (Royal Autograph Version) (in Thai).

Banpot Napompeth, 'Background, Threat, and Distribution of Mimosa Pigra L. in Thailand', in G. Lamar Robert and Dale H. Habeck (eds.), *Mimosa Pigra Management, Proceedings of an International Symposium, February 22–26, 1982, Chiang Mai*, Corvallis, Oregon: International Plant Protection Center, 1983, pp. 15–26.

Bock, Carl, *Temples and Elephants: Travels in Siam in 1881–1882*, London: Sampson Low, Marston, Searle & Rivington, 1884; reprinted Singapore: Oxford University Press, 1986.

Bowring, Sir John, *The Kingdom and People of Siam*, London: Oxford University Press, 1969, Vol. 1.

Brown, Ian, *The Élite and the Economy in Siam c.1890–1920*, Singapore: Oxford University Press, 1988.

Bruguiere, Mgr., 'Letter dated Bangkok 1829', published in *Annales de l'Association de la Propagation de la Foi* [Annals of the Association of the Propagation of the Faith], 1831, Vol. 5.

'Burmese Invasions of Siam, Translated from the Hmannan Yazawin Dawgyi', in Nai Thien (trans.), *Selected Articles from the Siam Society Journal, Relationship with Burma—Part 1*, Bangkok: Siam Society, 1959.

Callicott, J. Baird and Ames, Roger T. (eds.), *Nature in Asian Traditions of Thought*, Albany, New York: State University of New York Press, 1989.

Campbell, Reginald, *Teak-Wallah*, London: Hodder & Stoughton, 1935; reprinted Singapore: Oxford University Press, 1986.

Carter, A. Cecil (ed.), *The Kingdom of Siam 1904*, New York: G. P. Putnam's Sons, 1904; reprinted Bangkok: Siam Society, 1988.

Center for Agricultural Statistics, *Agricultural Statistics of Thailand, Crop Year 1991/92*, Office of Agricultural Economics, Ministry of Agriculture and Cooperatives, Bangkok, 1992.

Center for Conservation Biology, *Rapid Assessment of Forest/Wildlife/River Ecology in Area Affected by Kaeng Sua Ten Dam*, Bangkok: Mahidol University, 1992.

Charnvit Kasetsiri, *The Rise of Ayutthaya*, Kuala Lumpur: Oxford University Press, 1976.

Chaumont, Mr le Chevalier de, *Relation de l'Ambassade à la cour du roi de Siam* [An Account of the Embassy to the Court of the King of Siam], reprinted Bangkok: Chalermnit Press, 1985.

Child, Jacob T., *The Pearl of Asia*, quoted in *Foreign Records of the Bangkok Period up to A.D. 1932*, Bangkok: Office of the Prime Minister, 1982.

Chit Phumisak, *Khwam Pen Ma Khong Kham Sayam, Thai, Lao le Khawm, Le Laksana Thang Sankhom Khong Chu Chon Chat* [The Origins of the Words Sayam, Thai, Lao, and Khom and Social Characteristics of These People], 2nd edn., Bangkok: Samnak Phim Duang Kamol, 1981.

Christensen, Scott R. and Areeya Boon-long, *Institutional Problems in Thai Water Management*, Bangkok: Thailand Development Research Institute, 1993.

Cirlot, T. E., *A Dictionary of Symbols*, New York: Barnes and Noble, 1993.

Collins, Brian and Dunne, Thomas, *Fluvial Geomorphology and River-gravel Mining*, Special Publication 98, Sacramento: California Department of Conservation, 1990.

'Conserve Water or Face the Consequences', *Bangkok Post*, 1 November 1992.

Crawfurd, John, *Journal of an Embassy to the Courts of Siam and Cochin China*, London, 1828; reprinted Kuala Lumpur and Singapore: Oxford University Press, 1967 and 1987.

Culture and Environment in Thailand: A Symposium Sponsored by the Siam Society, Bangkok: Siam Society, 1989.

Davis, Richard, *Muang Metaphysics: A Study of Northern Thai Myth and Ritual*, Studies in Thai Anthropology 1, Bangkok: Pandora, 1984.

Demand Side Management for Thailand's Electric Power System; Five-Year Master Plan, Prepared by the International Institute for Energy Conservation, Bangkok, 1991.

Earl, George Windsor, *The Eastern Seas, or Voyages and Adventures in the Indian Archipelago, in 1832, 1833, and 1834*, London: W. H. Allen, 1837; reprinted Singapore: Oxford University Press, 1971.

Electricity Generating Authority of Thailand (EGAT), *Bhumibol Dam and Hydropower Plant*, Bangkok, 1993.

———, *Post Impoundment Environmental Evaluation and Development Planning of the Bhumibol and Sirikit Projects*, Bangkok, 1987, Vol. 5.

———, *Sirikit Dam and Hydropower Plant*, Bangkok, 1993.

Flood, Thadeus and Flood, Chadin (eds., trans.), *The Dynastic Chronicles Bangkok Era, the First Reign: Chaophraya Thiphakorawong Edition*, Tokyo: Center for East Asian Cultural Studies, 1978, Vol. 1.

Gee, C. D., *Irrigation in Siam: Nature and Industry*, Bangkok: Ministry of Commerce and Communications, 1930.

Gerini, G. E., *Chulakantamangala: The Tonsure Ceremony as Performed in Siam*, Bangkok, 1895; reprinted Bangkok: Siam Society, 1976.

Gervaise, Nicolas, *The Natural and Political History of the Kingdom of Siam*, Paris: Claude Barbin, 1688; English edn., trans. John Villiers, Bangkok: White Lotus, 1989.

Hafner, James A., 'Riverine Commerce in Thailand: Tradition in Decline',

Journal of the Siam Society, Vol. 62, Pt. 2, 1974.

———, *Salt, Seasons and Sampans: Riverine Trade and Transport in Central Thailand*, Asian Studies Committee Occasional Papers Series No. 6, International Area Studies Program, University of Massachusetts at Amherst, 1980.

Hallett, Holt S., *A Thousand Miles on an Elephant in the Shan States*, Edinburgh, 1890; reprinted Bangkok: White Lotus, 1988.

Hughes, Ted, 'The Vintage of River Is Unending', *River*, London: Faber and Faber, 1983.

Hutterer, Karl L., Rambo, Terry A., and Lovelace, George, *Cultural Values and Human Ecology in Southeast Asia*, Ann Arbor: University of Michigan, 1985.

Ichimura, Shinichi (ed.), *Southeast Asia: Nature, Society and Development*, Honolulu: University Press of Hawaii, 1976.

Ingram, James C., *Economic Change in Thailand 1850–1970*, Stanford: Stanford University Press, 1971.

Ishii, Yoneo (ed.), *Thailand: A Rice-growing Society*, trans. Peter Hawkes and Stephanie Hawkes, Honolulu: University Press of Hawaii, 1978.

Kaempfer, Engelbert, *A Description of the Kingdom of Siam 1690*, reprinted Bangkok: White Orchid Press, 1987.

Kunstadter, Peter, *Regional Integration in Northern Thailand: An Introduction*, University of Washington, Asian Studies Association, 1970.

La Loubère, Simon de, *A New Historical Relation of the Kingdom of Siam*, London, 1693; reprinted Kuala Lumpur and Singapore: Oxford University Press, 1969 and 1986.

Lando, Richard P., 'The Spirits Aren't So Powerful Any More: Spirit Belief and Irrigation Organization in North Thailand', *Journal of the Siam Society*, Vol. 71, Pt. 1, 1983, pp. 121–47.

Le Moigne, G., Barghouti S., and Plusquellec, H. (eds.), *Dam Safety and the Environment*, World Bank Technical Paper No. 115, Washington, DC: World Bank, 1990.

'Letter from HRH Prince Damrong Rajanubhab to HRH Prince Narisara Nuwattiwong of 5 November 1940', in *San Somdet* [Letters of Princes], Bangkok: Khurusapha Press, 1962, Vol. 20.

Lewis, Paul and Lewis, Elaine, *Peoples of the Golden Triangle*, London: Thames and Hudson, 1984.

McCarthy, James, *Surveying and Exploring in Siam*, London: John Murray, 1900.

McKinnon, John and Wanat Bhruksasri, *Highlanders of Thailand*, Kuala Lumpur: Oxford University Press, 1983.

Malloch, D. E., 'Extracts from My Private Journal Relative to Events Which Happened after the Departure of Captain Burney in 1826, to the Period of My Quitting Bangkok on the 20th March 1827', *The Burney Papers*, Vol. 2, Pt. 4, n.d.

Manas Chitakasem, 'The Emergence and Development of the Nirat Genre in Thai Poetry', *Journal of the Siam Society*, Vol. 60, Pt. 2, July 1972, pp. 135–81.

Milner, G. B., *Natural Symbols in Southeast Asia*, London: School of Oriental and African Studies, University of London, 1978.

Morphological Study of the Effects of Sand Dredging in the Chao Phya River, Thailand, Research Report No. 96, Bangkok: Asian Institute of Technology, 1980.

Olivares, Jose, 'Health Impacts of Irrigation Projects', in G. Le Moigne,

S. Barghouti, and H. Plusquellec (eds.), *Dam Safety and the Environment*, World Bank Technical Paper No. 115, Washington, DC: World Bank, 1990, pp. 149–64.

Oron, G. et al., 'Wastewater Disposal by Sub-surface Trickle Irrigation', *Water Science Technology*, Vol. 23, 1991, pp. 2149–58.

Oron, G. et al., 'Effect of Effluent Quality and Application Method on Agricultural Productivity and Environmental Control', *Water Science Technology*, Vol. 26, Nos. 7–8, 1992, pp. 1593–601.

'Out on a Limb', *Manager*, Bangkok, January 1992.

Pallegoix, J.-B., *Description du royaume Thai ou Siam* [A Description of the Kingdom of Thailand or Siam], reprinted Bangkok: D. K. Bookhouse, 1976.

Peltier, Anatole-Roger, *Pathamamulamuli: The Origin of the World in the Lan Na Tradition*, privately published, Chiang Mai, 1991.

Phra Ratcha Phongsawadan Krung Si Ayutthaya chabap Phra Paramanuchit [The Phra Paramanuchit Version of the Royal Chronicles of the Ayutthaya Dynasty], Bangkok: Ongkankha Khong Khrusapha, 1961, Vol. I.

Piyanart Bunnag, Duangporn Nopkhun, and Suwattana Thadaniti, *Khlong Nai Krungthep* [Canals in Bangkok], Bangkok: Chulalongkorn University, 1982.

Piyasena, W. and Nutalaya, P., 'Rehabilitation of a Depleted Aquifer System of Bangkok through Artificial Recharge', in *Soil, Geology and Landforms: Impact on Land Use Planning in Developing Countries. First International Symposium on Soil Geology and Landforms, Bangkok, 1–3 April 1982*, 1983, pp. D9.1–13.

Potter, Jack M., *Thai Peasant Social Structure*, Chicago: University of Chicago Press, 1976.

Prasoet Churatana, *Nan Chronicles* (trans., David K. Wyatt), Data Paper 59, Southeast Asian Program, Ithaca: Department of Asian Studies, Cornell University, 1966.

Prevost, A-F., *Histoire Generale Des Voyages* [General History of His Voyages], Paris, 1751, Vol. IX.

Prinya Nutalaya and Rau, Jon L., 'Bangkok: The Sinking Metropolis', *Episodes*, Vol. 1981, No. 4, 1981, pp. 3–8.

Project Completion Report of the First Inland Waterway Project, Bangkok: United Nations, 1990.

Purchas, Samuel, *Purchas: His Pilgrimage or Relations of the World and the Religions*, 3rd edn., London, 1617.

Rigg, Jonathan (ed.), *The Gift of Water: Water Management, Cosmology and the State in South East Asia*, London: School of Oriental and African Studies, University of London, 1992.

Roberts, Edmund, *Embassy to the Eastern Courts of Cochin China, Siam and Muscat during the Years 1832-3-4*, New York: Harper and Brothers, 1937.

Royal Irrigation Department (RID), *The Greater Chao Phya Project*, Ministry of Agriculture, 1957.

———, *Water Management for Rice Cultivation in Thailand*, Ministry of Agriculture, 1991.

Ruangdej Srivardhana, *No Easy Management: Irrigation Development in the Chao Phya Basin, Thailand*, Reprint No. 63, Honolulu: East–West Environment and Policy Institute, 1984.

Rungsi Prachonpachanuk, 'Bangkok under Water', *Bangkok Standard*, 31 August 1969, pp. 9–11.

Ruschenberger, W. S. W., *A Voyage around the World including an Embassy to Muscat and Siam 1835, 1836 and 1837*, Philadelphia, 1838; quoted in *Foreign Records of the Bangkok Period up to A.D. 1932*, Bangkok: Office of the Prime Minister, 1982.

Sacha Sethaputra, Theodore Panayotou, and Vute Wangwacharakul, *Water Shortages: Managing Demand to Expand Supply*, Research Report No. 3, 1990 TDRI (Thailand Development Research Institute) Year-end Conference, Jomtien, Bangkok: TDRI, 1990.

'Sanan: No Need to Get Tough with Factories', *Bangkok Post*, 29 May 1993.

Sathien Koses (pen name of Phya Anuman Rajadhon), 'Ban Tuk Chu, Le Chanit Rua Tang Khong Thai' [Descriptions of Names and Types of Thai Boats], in *Fuang Khwam Lang* [Recollections], Bangkok: Suksit Siam Press, 1968, Vol. 2, pp. 246–9.

Seni Pramoj, MR and Kukrit Pramoj, MR, *A King of Siam Speaks*, Bangkok: Siam Society, 1987.

Singaravelu, S., 'The Legend of the Naga-princess in South India and Southeast Asia', Paper presented to the 25th Session of the All-India Oriental Conference, Calcutta, 29–31 October 1969; reprinted in Tej Bunnag and Michael Smithies (eds.), *In Memoriam: Phya Anuman Rajadhon. Contributions in Memory of the Late President of the Siam Society*, Bangkok: Siam Society, 1970, pp. 9–15.

Small, Leslie, 'An Economic Evaluation of Water Control in the Northern Region of the Greater Chao Phya Project of Thailand', Research paper, South East Asia Development Advisory Group, Washington, DC, 1971.

Solheim, Wilhelm G., 'The New Look of Southeast Asian Prehistory', *Journal of the Siam Society*, Vol. 60, Pt. 1, January 1972, pp. 1–20.

Somboon, J. R. P., 'Coastal Geomorphic Response to Future Sea-level Rise and Its Implication for the Low-lying Areas of Bangkok Metropolis', *Tonan Ajia Kenkyu* [South-East Asian Studies], Tokyo, Vol. 28, No. 2, September 1990, pp. 155–68.

Somboon, J. R. P. and Thiramongkol, N., 'Holocene Highstand Shoreline of the Chao Phraya Delta, Thailand', *Journal of Southeast Asian Earth Sciences*, Vol. 7, No. 1, 1992.

Sommai Premchit and Amphay Dore, *The Lan Na Twelve-month Traditions*, Chiang Mai: Toyota Foundation, 1992.

Sommerville, Maxwell, *Siam on the Meinam; From the Gulf to Ayuthia*, London: Sampson Low, Marston and Company, 1897; reprinted Bangkok: White Lotus, 1985.

Stott, P. A., *Nature and Man in South East Asia*, London: School of Oriental and African Studies, 1978.

Sumet Jumsai, *Naga: Cultural Origins in Siam and the West Pacific*, Singapore: Oxford University Press, 1988.

Tachard, Guy, *Voyage to Siam*, London, 1688; reprinted Bangkok: White Orchid Press, 1981.

Terwiel, B. J., *Through Travellers' Eyes: An Approach to Early Nineteenth Century Thai History*, Bangkok: Editions Duang Kamol, 1989.

Thailand Country Report to the United Nations Conference on Environment and Development (UNCED), June 1992, Ministry of Science, Technology and Industry, Bangkok, 1992.

Thaweesak Jaritkhuan, 'A Study of Distribution in Sand Sediment Layers on the Bank of the Old Chao Phya River in Phra Nakhon Si Ayutthaya', M.Ed. thesis, Srinakarinviroj University (Prasanmitr), Bangkok, 1990.

'The Official Answer: Dam Projects', *Bangkok Post*, 3 November 1993.

Thiva Supajanya, 'Sukhothai: Its Hydraulic Past', *Geology*, Bangkok, Vol. 15, No. 1, 1992.

Uraivan Tak-kim-yong, *The Karen Culture: A Co-existence of Two Forest Conservation Systems*, Chiang Mai: Faculty of Social Sciences, Chiang Mai University, 1990.

———, *Natural Resource Utilization and Management in the Mae Khan Basin: Intermediate Zone Crisis*, Chiang Mai: Faculty of Social Sciences, Chiang Mai University, 1988.

Van der Meer, C. L. J., *Rural Development in Northern Thailand: An Interpretation and Analysis*, Groningen: Krips Repro Meppel, University of Groningen, 1981.

Van Liere, W. J., 'Traditional Water Management in the Lower Mekong Basin', *World Archeology*, Vol. 11, No. 3, February 1980, pp. 265–88.

Vanpen Surarerks, *Historical Development and Management of Irrigation Systems in Northern Thailand*, Chiang Mai: Department of Geography, Chiang Mai University, 1986.

Van Ravesswaay, L. F., 'Jeremias Van Vliet, Description of the Kingdom of Siam', *Journal of the Siam Society*, 1910, pp. 25–6.

Wales, H. G. Quaritch, *Siamese State Ceremonies*, London: Bernard Quaritch, 1931.

Wijeyewardene, Gehan, 'A Note on Irrigation and Agriculture in a North Thai Village', *Felicitation Volumes of Southeast-Asian Studies, Presented to His Highness Prince Dhaninivat Kromamun Bidyalabh Bridhyakorn on the Occasion of His Eightieth Birthday*, Bangkok: Siam Society, 1965, Vol. II, pp. 255–9.

Winai Pongsripian, 'A Historian's Comment', *Geology*, Bangkok, Vol. 15, No. 1, 1992.

Workshop on Uses and Losses due to Mimosa Pigra, Chiang Mai: Chiang Mai University, December 1986.

Wyatt, David K., *Thailand: A Short History*, Bangkok: Yale University Press/Thai Wattana Panich, 1984.

Young, Ernest, *The Kingdom of the Yellow Robe: Being Sketches of the Domestic and Religious Rites and Ceremonies of the Siamese*, London: Archibald Constable & Co., 1898; reprinted Kuala Lumpur: Oxford University Press, 1982.

Index

Artificial Rain-making, 192
Artificial recharge, see Groundwater recharge
Ayutthaya: canal engineering in, 41, 121; city canals in, 34; foreigners in, 32, 34–6, 41; in history, 34–6; influence in Bangkok's design of, 45; river at, 4, 10–12, 34, 121, 156; river in commerce of, 32–4, 39, 41–2; river engineering in, 39, 121; river in history of, xiv, 19, 29, 31–2, 42; river in warfare of, 29, 31–2, 42
Ayutthayan: agriculture, 73; houseboats, 34, 55; shipbuilding, 34, 47
Ayutthayan Chronicles, 6, 29, 34, 41
Ayutthayan stone pillar, 135, 137

Bang Li Houses, 72
Bangkok: canals, 44–7, 49, 53, 62, 121, 187; city canal engineering, 121–4; environmental problems and remedies, 162–4, 168, 188–9; floods, 5, 44, 62, 127, 128, 129, 135, 137, 145, 146; fortifications, 48, 50; houseboats, 55, 56, 61, 63, 169; period canal engineering, rural, 48–50, 53–4, 121–3, 128; ports, 36, 39, 40, 163; river at, 4, 7, 56; river in commerce, 47, 50, 53–5, 57, 60, 63, 90; river in development of, 47, 62, 66; river in history of, 39, 40, 135; river in role of, 44; river in warfare, 41, 47–8; role in river commerce, 41; subsidence, 165–7; water shortages, 148, 150, 172
Bangpakong River: Buddha images, 113; canals, 40, 49, 123; in history, 10; proposed development of, 195
Barge processions, Royal, 81, 83, 110–12
Barges: Ayutthaya period, 80, 81; Bangkok period, 45, 49; commercial, 57, 62, 63, 88–9, 187; royal (Ayutthaya period), 38; royal (Bangkok period), 48, 81–3, 87; royal replicas, 99; warfare, Ayutthayan, 80

Bhumibol, King, 28, 100, 112, 140, 189, 192
Bhumibol Dam, 1, 3, 23, 78, 83, 90, 140, 144–5, 150–3, 178, 192, 194
Bight of Bangkok, 4, 18
Boatbuilding, 47
Boats: in Bangkok, 45; in commerce, 23, 53, 55, 57–8, 61, 63, 74, 90, 134, see also Storeboats; construction techniques, 86, 89; decline in importance, 62–6, 90; in fishing, 75, 78, 86–8; Lanna period, 23; in literature, 117–18; new uses, 91–2, 187–8; prehistoric, 8; processions, 46, 48, 82–3, 111–13, see also Barge processions; spirits of, 83, 109, 113; terms, 83; Thai facility in paddling, 80; use in Ayutthaya, 34, 38; use in daily life, 79–80, 83, 86–9, 114, 116, 134; in warfare, 32, 42, 48, 80–3, 126; see also Houseboats
Boat types: royal boats, 83; *rua khut*, 85–6; *rua mat serm krap*, 85–7; *rua taw*, 85, 87–9
Bronson, Bennett, 18
Buddha: and rivers, 71; and water, 96
Buddhism: and rivers, 10, 45, 50, 55, 71, 79, 109, 112–13; and water, 94, 97–8, 100–5, 107, 110, 115
Buddhist monk ordination, 101, 104
Burma: peoples of, 13, 18, 48; spirit belief, 95, 108; timber industry, 58; wars with, 32, 40, 42, 44, 48, 80, 100
Burmese: geography, 1, 14, 31; rites and festivals, 96, 99, 110; river diversions, 195–6; temples and buildings, 59

Canal: Department, 126–7; dredging rites, 17; excavation proposals, 195; maintenance, 154; pollution and remedies, 162–3, 188–9
Canal types: *khlong chuam maenam*, 38–40, 121; *khlong lat*, 38–9, 49, 121, 154; *khlong rop muang*, 38, 44–5

Canals: Ayutthaya, 29, 34, 38, 39; Bangkok, 45–7, 54, 62–3, 92; hilltribe, 17; Lanna period, 16, 22; Phra Ruang Road, 27–8; prehistoric, 10, 18; rural: Ayutthaya period, 40–1, 48, Bangkok period, 11–12, 48–9, 50, 53–4, 121–9, 135, 137–9, 141–4, 150, 153, 167, 176, 181, 183; Sukhothai period, 27–8; Thonburi, 44; U-Thong, 10–11
Canals dug during reign of: Rama II, 48–9; Rama III, 49–50, 121; Rama IV, 53–4, 121; Rama V, 122–3, see also Rangsit Project; Rama VI, 128; Rama IX, see Greater Chao Phya River Project
Chainat, 4, 10, 11, 29, 37, 63, 72, 137, 139, 150, 187, 189
Chainat Dam, see Chao Phya Dam
Chalawan, 95, 116
Chao Phya River: in agriculture, 24, 31, 73, 74, 141, 143, 175; boats, 87–8, 90; bridges, 65; in commerce, 32–3, 47, 50, 53–61; course of, xiii, 3–4, 9, 10, 11, 12; in culture, 110, 112, 113; dams, 11, 65, 126, 138, 187; derivation of name of, 5–6, 13; development proposals, 192, 194, 195, 196; engineering: Ayutthaya period, 11, 31, 38–41, Bangkok period, 12, 47–9, 121, 122, 137–9, 142; environmental problems, 66, 146, 154–5, 161, 167, 172, 188; floods, 25, 135, 145–6; foreign mention of, 6, 18, 32, 33, 41, 50, 53, 55, 56, 61, 156; fortifications: Ayutthaya period, 34, 41, 44, Bangkok period, 48, 50; geological history of, 7–8, 10–11; historical role of, xiv, 19, 29, 31, 32, 39, 40, 42, 45, 135; original five channels of, 11; ports, 187; regimes, 5; role in cultural dissemination, 57; role in warfare: Ayutthaya period, 29, 31–3, 42, Bangkok period, 41, 44, 47–8; settlement patterns, 68; shipping, 33
Chao Phya Dam, 65, 126–7, 137–9, 144–5, 161, 195

207

INDEX

Chao Phya Noi, 4, 9, 11–12
Chiang Mai: environmental problems and remedies, 156–9, 168, 189–90, 192; floods, 146; foreign mention of, 37; Ping's role in history of, xiv, 21–2, 31–2, 36, 63–4, 69, 90, 129, 146; river at, 1; in river culture, 98, 101; in water culture, 97
Chinese: canal financiers, 123; canal labourers, 49, 53, 123–5; riverside merchants, 47, 55, 58; shipbuilders, 34, 47, 83
Chinese-design boats, 87, 90
Chulalongkorn, King, *see* Rama V
Compensation dams, 179
Crawfurd, John, 4, 6, 50, 55, 58, 136
Crocodiles, 95, 107, 114, 117–19, 135

Dam: construction, opposition to, 183, 193, 194; projects, results of, 141–3
Deforestation, xv, 59, 129, 131, 156–8, 160, 169–70
Department of Mineral Resources, 164–5
Diseases: water-borne, 99, 110, 136, 145–6, 163
Ditches and dikes programme, 140–2
Dredging: Ayutthaya period, 39; Bangkok, 62; environmental problems, xv, 158, 161, 166; Thonburi period, 44
Dynastic Chronicles: First Reign, 45, 46, 135; Second Reign, 48, 50

Embayment of Central Plains, 18
Environment Protection Agency, 156
Environmental Fund, 188

Ferry Boats, 66, 86, 187
Festivals, water-related, xiv, 96, 98–100
Fish: to destroy pests, 189–90; destruction of, xv, 46, 156, 161, 163, 182; fireworks, 114; in literature, 116–19; products, 53, 55, 58, 75, 113; varieties, 13, 74, 108, 161, 189
Fishing: boats used in, 86–8; future of, 155; history of, 13, 25, 135; prohibitions against, 78–9, 110; techniques of, 63, 69, 74, 75, 76, 78
Floating communities, *see* Houseboats
Floating markets, 61, 74
Flood control schemes, 126, 130, 135–7, 140, 144, 167, 178, 194
Flood plain agriculture, 18, 24, 28, 72, 74, 142, 185
Floods: and architecture, 72; Ayutthaya period, 35; Bangkok period, 44, 62, 127–9, 135, 137, 145–6; Chiang Mai, 21; in culture, 96, 107; and disease, 146
Food and Agriculture Organization: predictions, 155; study, 137, 183, 191

Gervaise, Nicolas, 6, 13, 33, 34, 36, 37, 38, 41, 156
Gods: earth or crop-related, 104, 106; river-related, 98, 108; water-related, 95, 96, 97, 100, 103, 105, 109
Greater Chao Phya River Project, 129, 135, 137–44, 150, 153, 167, 176, 181, 183
Groundwater Act, 165
Groundwater contamination, 162
Groundwater extraction, 164–5, 178, 184–5, 196
Groundwater recharge, 167
Gulf of Thailand: geological development of, 12; in prehistory, 8, 10, 11, 12; and rivers, xiii, 5, 39, 40, 48, 166

Hilltribe: agriculture, xiv, 17, 68, 157, 169, 192; beliefs, 69, 95, 106; use of water, 68, 69, 95, 106, 192
Houseboats: Ayutthaya, 34, 55; Bangkok, 55, 56, 61, 63, 169; Nan River, 63; Phitsanulok, 63
Hua na fai, 16, 17, 132–3
Hydroponics, 74, 160

Ing River, 7, 14, 195
Ingram, James, 128
Irrigation: efficiency, 182–4, 193; systems, Chinese antecedents, 13
Irrigation Act: National, 130, 181, 183; People's, 130

Kaeng Soi Rapids, 23, 83, 90
Kaeng Sua Ten, 3, 90; dam proposal, 180, 192, 194, 195; environmental report, 194
Kamphaeng Phet, 7, 26, 27, 28, 33, 36, 86, 158
Karen: beliefs about water, 106; conservation practices, 169, 192
Khiu Lom Dam, 141, 146, 159
Khlong Bangkok Noi, 39, 44, 62, 82, 112, 187
Khlong Bangkok Yai, 39, 44, 53, 62, 78, 112, 122
Khlong Banglamphu, 45, 53, 62
Khlong Chao Chet, 72
Khlong Lawd, 44, 45, 47, 53, 62
Khlong Mahanak, 45, 46, 49, 62, 114
Khlong Ong Ang, 45, 53, 62
Khlong Padung Krung Kasem, 53, 62
Khlong Sansaep, 49, 62, 63, 92
Khlong Toey: port, 36, 40, 163; port pollution, 163
Khmer: irrigation techniques, 18, 24, 25, 186; water spirits, 95, 96
Kings and irrigation, 5, 16, 21, 25–8, 124, 125, 126, 127, 128, 189
Kok River, 7, 195
Kraithong, 116

La Loubère, Simon De, 6, 33, 34, 37, 38, 40, 81, 189
Lahu beliefs about water, 106
Lampang, 3, 36, 59, 64, 69, 141, 158
Lamphun, 97, 99, 129, 158
Lanna Chronicles, 14, 21, 23, 110
Lisu beliefs about water, 69, 95
Logging, 3, 57–60, 160, 182
Lopburi, 4, 29, 32, 36, 40, 193
Lopburi River, 4, 9, 12, 31, 137
Loy Krathong, xiv, 98–9, 188

Mae Khongkha, 98, 99, 108
Mae Klong River, 44, 48, 180; diversion, 195
Mae Prosop, 104
Mae Torani, 96, 100
Mae Ya Nang, 83, 109
Mekong River, 7, 14, 31, 73, 195
Mengrai, King, 16, 21, 22, 97, 101
Mengraisat laws, 16, 23
Metropolitan Waterworks Authority, 163, 164, 178
Mien beliefs about water, 106
Mimosa pigra: eradication, 159–60; infestation, 158–9
Mongkut, King, *see* Rama IV
Muang fai: administration, 133, 169, 177; associations, 130, 132, 140; conflicts, 131, 177; contracts, 16, 130, 131, 169; fees, 129, 132, 183; laws, 16, 17, 131; maintenance, 16; spirits, 17, 169, 170; system, 14–17, 19, 26, 72, 73, 130–4, 144, 177, 183
Musur, *see* Lahu

Nak (*naga*), 70, 95–7, 101, 104, 105, 113
Nakhon Sawan: festivals, 96; in history, 37, 65; and logging, 59–60; proposed port, 187; and rivers, xiii, 3, 4, 8, 10, 37, 63, 90, 152, 163, 187; and war, 31, 48
Nan River: agriculture, 74, 141; boats, 90; course of, xiii, 3, 4; in culture, 113, 116; derivation of name of, 13–14; dams, 65, 141; development and proposals, 191, 193–5; environmental problems, 189;

INDEX

houseboats, 63; pollution, 188; proposed port, 187; rapids, 90
Nan town, 31
Narai, King, 5, 34, 42, 81
National Environment Board, 156, 179, 188, 189
National Potable Water Scheme, 164
National Water Resources Committee, 179
Nature symbolism, 94, 95, 105, 106, 107, 114, 115
Nguak, 95
Nid Hinshiranan, 39

OK TAEK TOWNS, 28, 29

PAKNAMPHO, *see* Nakhon Sawan
Pasak Dam, *see* Rama VII Dam
Pasak River: description of, 4, 31, 88, 139, 187; proposed development of, 193, 195; scheme, 126–8
Phitsanulok, 4, 8, 28, 36, 193; houseboats, 63, 169
Phra Ruang Road, 26, 27, 28
Phrae: in history, 3, 31, 59, 64, 69, 90, 169, 192; Shan Rebellion, 126
Phya Anuman Rajadhon, 72, 84, 85, 99, 108, 148, 169
Ping Kao, 21, 129
Ping River: boats, 83, 90; bridges, 64; commercial role of, xiv, 90; course of, xiii, 1, 3, 4; in culture, 98, 99, 100, 104, 116; dams, 65, 129, 130, 140, 141; derivation of name of, 14; development and proposals, 178, 187, 191, 195; environmental problems, 157, 159, 168; floods, 146; historical role of, 21, 22, 23, 26, 27, 32, 48; and Phra Ruang Road, 27; rapids, 23, 83, 90
Ports: Khlong Toey, 36, 163; proposed, 187; Thonburi customs, 41
Pump irrigation, 17, 133, 145, 177, 181, 185

RAILWAYS AND RIVERS, 57, 63, 65, 126, 127, 128
Rama I's reign: barge construction in, 81; canal projects in, 45, 114; construction of Bangkok in, 45–6; river fortifications in, 48
Rama II's reign: barge construction in, 81; canal construction in, 48–9; river engineering in, 11; river fortifications in, 48
Rama III's reign: Ayutthayan stone pillar, 135; barge construction in, 81; canal projects in, 49; river fortifications in, 50

Rama IV's (King Mongkut) reign: canal projects in, 53, 54, 121–3; comments on floating towns in, 55; comments on foreign ignorance of river name in, 7; comments on river sanitation in, 162; royal barges in, 112
Rama V's (King Chulalongkorn) reign: barge construction in, 82; canal projects in, 74, 123; and Phra Ruang Road, 26; railway projects in, 126; water ceremonies in, 99, 114; water supply in, 164
Rama VI's (King Vajiravudh) reign: canal projects in, 127; royal barges in, 82
Rama VI Bridge, 60, 65
Rama VII Dam, 128, 137
Ramathibodi II, river engineering, 39
Ramkamhaeng, King, 21, 25, 98, 110
Rangsit Project, 124–8, 137, 181
Reforestation, 191
Rice: boats, 86, 87, 89; and canal construction, 49, 123, 124; cultivation techniques, 18, 24–5, 28, 72–4, 106, 143, 174; and floods, 18, 24, 127, 128, 135, 137, 146, 189; new approaches, 176, 178, 183, 189, 191; and river engineering, 127, 129, 131, 137; trade: domestic, 57, 61, 65, foreign, 31, 32, 33, 47, 53, 74, 123, 127, 137, 146, 174, 175, 178; varieties: dryland, 14, 17, 69, 72, 73, 143, 174, floating, 72, wetland, 13, 18, 24, 31, 72, 73, 143; water requirements for, 73, 137, 143, 175; and water shortages, 149, 154, 174, 175, 176; yields, 73, 142, 143, 174, compared with world average, 191
Rites: Buddhist, 50, 101, 103, 104; hilltribe, 106; loyalty oaths, 100, 101; *muang fai*, 16–17, 104, 169; of passage: ordinary, 100, 102, royal, 100, 101; to propitiate river spirits, 109; to protect rivers, 99, 169; rain-calling, 96, 101, 104, 105, 106; river-related, 109; using river water, 100; village, 98, 104, 105; water-related, 97
Rivers: children's games, 118; lengths, 1, 3, 4; proverbs, 117; songs, 116; symbolism, xiv, 22, 106, 107, 108; and town design, 28, 29, 31
Royal Irrigation Department (RID): and Bhumibol Dam, 140; and Central Plains irrigation, 144–5, 154; and Ditches and Dikes Act, 140; and the Greater Chao Phya, 128, 137–8; and *muang fai* system, 17; and northern irrigation, 129–30, 132;

predecessors, 126, 127, 129; proposals, 177, 192, 193, 194, 195; and user fees, 183; and Users' Associations, 130, 183; and water conflicts, 150; and water sanitation, 164; and water shortages, 150, 152, 153, 154, 175, 178, 192; and water weeds, 159, 160; yield projections, 142

SALWEEN RIVER, 31; diversion of, 195, 196
Samut Prakan, 4, 6, 12, 48, 50, 87, 112–13, 136
Saritphong Dam, 22, 25, 137
Science, Technology, and Environment Ministry, 196
Shan: people, 69; Rebellion, 126; States, 13, 14, 104
Sirikit Dam, 4, 63, 78, 140, 141, 144–5, 150–4, 178, 187, 194, 195
Si Satchanalai, 3, 29, 31; and Phra Ruang Road, 26, 27, 28
Slash-and-burn agriculture, *see* Swidden agriculture
Songkran, xiv, 96, 97, 105, 108
Spirits: boat, 83, 109, 113; earth, 70; hilltribe, 69, 95, 106; *muang fai*, 17, 169, 170; river, 108; water, 110
Steam launches, 90
Storeboats, 57, 58, 63, 90, 91
Sugar: and canal construction, 121–3; trade, 53, 121, 123
Sukhothai period: irrigation, 25; water engineering, 25
Suphan Buri: irrigation scheme, 127, 128; river, *see* Tha Chin River; town, 28, 29, 32, 36; water conflicts, 150
Swidden agriculture, xiv, 17, 68, 157

TABOOS: reduced power of, 169
Tak, 3, 5, 23, 32, 36, 42, 48, 59, 64, 83, 158, 195
Taksin, King, 44, 45, 47
Taphan Hin proposed port, 187
Tha Chin River: Buddha images, 113; canals, 40, 48, 121, 122; description of, 4, 9, 10, 33, 88, 122; flooding, 146; irrigation scheme, 139; river proposal, 195
Thonburi: canals, 44; in history, xiv, 4, 36, 39, 41, 44–5, 47, 62, 74, 91, 112, 166

U-THONG, PRINCE, 29, 31
U-Thong canals, 10

VAJIRAVUDH, KING, *see* Rama VI
Van der Heide, Homan, 126–7, 137, 183

Van Vliet, Jeremias, 111
Village designs: hilltribe, 68; riverine, 69, 70

WANG RIVER: commercial role of, 59; course of, xiii, 3; in culture, 104, 116; dams, 65, 146; development, 154, 191; derivation of name of, 14; environmental problems, 158, 160; foreign mention of, 37; pollution, 155
Ward, Thomas, 127
Wastewater: discharges, 165, 169, 185, 188; treatment, 162, 164, 167, 188–9, 195
Wat Phra Chedi Klang Nam, 50
Wat Saohai manuscript, 17
Water: and creation theories, 95; hyacinth, 168; pollution and new remedies, 148, 150, 154–6, 161–3, 169, 182, 184, 188, 189–90; shortages and solutions, 142–3, 146, 148–50, 153–4, 165, 172, 174–6, 178, 183, 192, 195; symbolism, xiv, 13, 94–6, 98, 102, 104, 106–7; user associations, 130, 133, 143; user fees: rural, 129, 132, 181, 183, 184, 185, urban, 184, 185; user groups, 130, 143; users associations, 177; vegetation, 74, 158, 159, 160, 168
Water Law, 181, 195
Water Ministry, 181
Watershed: destruction, 130, 131, 155–7, 169, 182, 191–2; restoration, 191, 194
Water-wheels, 133
Wiang Kum Kham, 21
Wiang Soi Si Suk, 23
Wilaiwan Khanittanan, 95, 115
World Bank, 137, 140, 182, 187, 189, 194
Wyatt, David K., 32

YAO, *see* Mien
Yom River: commercial role of, 59, 90; course of, xiii, 3, 4; in culture, 98, 116; derivation of name of, 14; development and proposals, 141, 154, 180, 187, 191, 192, 194, 195; engineering, 25; environmental problems, 194; foreign mention of, 36, 37; historical role of, xiv, 22, 24–8, 36, 90; and Phra Ruang Road, 27; pollution, 158; rapids, 90
Yue irrigation systems, 13, 14